CRUSADERS FOR WILDLIFE

A History of Wildlife Stewardship in Southwestern Colorado

Glen A. Hinshaw

WESTERN REFLECTIONS
PUBLISHING COMPANY
OURAY, COLORADO
2000

First edition

Library of Congress Catalog Number 00-102171

ISBN 1-890437-51-4

Cover and book design by Paulette Livers Lambert

Western Reflections Publishing Company
P.O. Box 710
Ouray, Colorado 81427
USA

CRUSADERS FOR WILDLIFE

To Margaret
Best Wishes,
Glen A. Hinshaw

This book is dedicated to my wife Carol for lovingly putting up with me during my writing sessions.

I also dedicate this book to the wonderful citizens of Creede, Colorado, among whom I had the privilege of living.

This book is also dedicated to the Game Wardens, Wildlife Conservation Officers, District Wildlife Managers, supervisors, wildlife technicians, and biologists listed in the Appendix, who have protected wildlife and served the public in the Upper Rio Grande and to all my coworkers in the Colorado Division of Wildlife.

CONTENTS

ACKNOWLEDGMENTS

I wish to thank the following for their contributions to this book:

Conrad Albert, John Alves, Jerry Apker, Jim Basham, Joyce Becker, Ed Bechaver, Don Benson, Norman Bishop, Fred Blackburn, Ray Boyd, Richard & Dianna Brown, Ruth & Darcy Brown, Lucille Buchanan, Gene Byrne, Helen Caughlan, Dick Cochran, Jim Cochran, Cliff Coghill, Dave Croonquist, Dena M. Dority, Charles Downing, Dick Fentzlaff, Ed Dumph, Carol Ann Wetherill Getz, Floyd Getz, Paul Gilbert, Glenn Gross, Ed Hargraves, Steve Hartvigsen, Bill Haggerty, Lloyd Hazzard, Carol Hinshaw, Esther Hosselkus Hinshaw, Judith Holmes, Margie Hosselkus, A. J. Hosselkus, Bob Hoover, Jim Houston, Janis Jacobs, Jeff Johnson, Bob Keiss, Howard Kennell, Dave Kenvin, Wayne Knisely, George Kroenert, Margaret Lamb, Dave Langlois, Tom Martin, Don Masden, Jim Mason, Mildred Mason, Phil Mason, Caroline Feast McCracken, Barry Nehring, Janice Nelson, Jim Olterman, Jerry Pacheco, Jerry Pearson, Jerry Poe, Tony Phipps, Dean Prentice, Laurence Riordon, Bob Rouse, Wayne Russell, Bill Rutherford, Dr. Ron Ryder, Jay Sarason, Gordon Saville, Mike Simpson, Don Smith, Randi Sue Smith, Howard Spear, Vince Spero, Tammy Fox Spezze, Charles and Dorothy Steele, Joel Swank, Bill Swinehart, Geoff Tischbein, Darryll Todd, Ivan Vanaken, Ron Velarde, Scott Wait, Robert Wardell, Jim Webb, Clayton Wetherill, Ernest Wilkinson, Dot Wilkerson, Bill Wiltzius, Ed Wintz, Rod Wintz, Brent Woodward, Harry Woodward, Jim Young, and Mike Zgainer, and the D. C. Booth National Fish Hatchery Museum, the Colorado Historical Society, and the Ute Indian Museum, Montrose, Colorado.

PREFACE

People who live, work, and recreate in the mountains of Colorado have a rich wildlife heritage. Many don't realize it, but some wild animals were already on the brink of extinction a hundred years ago. Now most species have been restored, and it is easy to take for granted the wild animals, birds, and fish that enhance our lives in what remains of these wild places in southwestern Colorado.

We are indebted to our Creator who gave us these beautiful mountains and valleys that will sustain us if we don't abuse them. We are also indebted to the Ute Indians who lived here for hundreds of years. They demonstrated that we are indeed a part of this ecosystem and that we, in spite of our concept of ownership, do not really own these mountains. Those who live in western Colorado merely possess title to lands that were confiscated from the Utes. We are here as briefly as the delicate blossoms of the tundra, when measured in San Juan Mountain time, yet we have left our mark upon the land.

The people who have lived in the San Juan Mountains for the past 100 years extracted their living from the local minerals, grass, timber, and wildlife and made many mistakes in managing resources. Our wildlife heritage was nearly lost through abuse of the once abundant wildlife and the natural habitat necessary for its survival. In some cases its purposeful extermination was the goal of the early settlers. Even though the natural environment has been fragmented and in some cases destroyed, natural processes such as regeneration and decay continue.

This book is a history of wildlife and wildlife management in and around the San Juan Mountains of southwestern Colorado. This book is also centered in the memories that people shared with me during my tenure as the wildlife officer in the Creede District of the Upper Rio Grande.

I have made every effort to keep facts and dates accurate, but some discrepancies may have crept in. Where possible, memories have been corroborated by newspaper accounts, reports, diaries, journals, and other people's recollections. I traveled thousands of miles to learn additional information, interviewed dozens of people, and gathered letters, photographs, journals, and diaries. I've tried to spell names correctly and quote what people told me. In some cases quotes have been paraphrased and edited to clarify meaning. Some memories could not be corroborated, however, and there may be some mistakes with dates. Occasionally I found cases of people remembering the same event differently. Individual prejudices and perspectives also filter the conclusions and feelings that people have.

This book describes a crusade for wildlife by hundreds of people who carried out many wildlife projects, programs, and efforts to perpetuate wildlife and people's enjoyment of it. Since most of my career was spent in the Upper Rio Grande, many of the examples and details are from that area, but they reflect what was happening in many western communities. Faceless agencies and organizations have played an important role, but where appropriate I've used individual names. A great part of the crusade was performed by people who, for the most part have never been recognized for their contributions. We owe them honor and tribute. Without some appreciation of this legacy, we are poorer. Recognizing our debt does not make us poorer, but rather makes us richer with knowledge. Learning the history of our wildlife heritage and the legacy of stewardship that brought it back from near extinction brings a responsibility. We must pass on what we have learned to our youth as well as our thousands of visitors. Without understanding and gratitude we remain dreadfully poor. A grateful heart is a happy one and I hope that this book will make the reader smile in appreciation. I have attempted to share these memories in a context of what was happening to and for wildlife nationally, statewide, and locally. You will meet unfamiliar people who played an important role in perpetuating the wildlife heritage in southwestern Colorado. There are people just like them in nearly every western community. You may also find familiar names, and will be reminded of things they did to preserve our wildlife. The one thing that I couldn't get into print were the voices of the past: the giggles, the anger, the laughter, the sadness, and the joy of people remembering their experiences with the wildlife and a few "wild" people.

CRUSADERS FOR WILDLIFE

CHAPTER I

ABUNDANCE TO EXPLOITATION

The Country

The rugged San Juan Mountains contain more country above 13,000 feet than any other place in the United States. The western slope of the San Juan high country has numerous vertical faces and pinnacles with dozens of clear blue alpine lakes nestled in its cirques. The many "fourteener" peaks reach to the blue sky with names like the Needles, Wilson, Handies, Eolus, Uncompahgre, Wetterhorn, Sneffels, and San Luis.

The Rio Grande side of the Continental Divide, although just as high and rugged, has more rounded mountains. The Continental Divide at 12,588-foot Stony Pass is the farthest point west that the spine of our nation extends in Colorado. Here the San Juans wrap like a horseshoe around the headwaters of the Rio Grande. The mighty Rio Grande Pyramid which reaches 13,873 feet is the sentinel of the headwaters while the La Garita Mountains form the northern ridge of this huge basin.

The high Divide country is carpeted with beautiful tundra flowers, alpine willow, and potholes edged with sedge marshes. Only a few passes dip down to the timberline. The Utes traveled through these gateways which now have names like Weminuche, Piedra, San Luis, Spring Creek, and Wolf Creek, to their hunting grounds in the San Juan and LaGarita Mountains and the broad San Luis Valley.

Below the alpine zone there are stands of spruce, fir, and aspen that wild fires have made into a mosaic of forest and meadow. The San Juan Mountains cradle the headwaters of the San Miguel, which joins the Dolores River as it flows to the Colorado River. The Animas, Pine, and Piedra Rivers combine their tributaries to form the San Juan River that joins the Colorado River in what is now Lake Powell. The Lake Fork, Cochetopa, Tomichi, Cimarron, and Uncompahgre Rivers combine their waters with the East and Taylor Rivers to form the Gunnison River which in turn joins the Colorado River in the city of Grand Junction. The mighty Rio Grande begins as a spring on Stony Pass; plunges as pristine rapids and foam, gathering its tributaries into a majestic river which flows more than 1,800 miles to the Gulf of Mexico. Throughout this vast land the Ute Indians lived in harmony within the ecosystem, through natural cycles of abundance and famine directed by the climactic changes over the years.

When the ancient spruce forests of the San Juans were mere seedlings, only the Ute Indians were there to hunt and fish. Bighorn sheep, elk, deer, buffalo, antelope, grizzlies, black bears, wolverines, lions, bobcats, pine martins, beavers, birds, fish and many other varieties of wildlife lived here in a dynamic equilibrium within the ecosystem.

Willow, alder, and narrow leaf cottonwood trees shaded the stream banks providing habitat for nesting birds. Hatching insects fell into the cold waters that teemed with Rio Grande or Colorado River cutthroat trout. River otter, beaver, mink, and muskrat lived in the beaver ponds and crystal clear streams. Tall grasses carpeted the meadows and mountainsides. Bald and golden eagles, hawks, and falcons nested in the cliffs and tall cottonwoods and soared above the meadows and waters searching for rodents, birds, fish, and other prey. However, the clear alpine lakes and high altitude streams that were above waterfalls were all fishless. The cutthroat trout could not get above the waterfalls to reach these upper stretches.

The Native Americans

The Upper Rio Grande was primarily the hunting ground of the Capote band of Ute Indians, who lived in the San Luis Valley and north-central New Mexico. The Weminuche band lived in the San Juan Basin but hunted on both sides of the San Juans and traveled major trails over Weminuche and Piedra passes and across the Upper Rio Grande to Spring Creek and San Luis passes. The Tabaguache Utes lived in the Uncompahgre and Gunnison drainages.

The Utes believed that the Creator provided the wild game for their people. What was given by the Creator was to be used, respected, and not wasted. Ute men spent much of their time hunting, fishing, and making the tools used in the pursuit and capture of game and fish.

Actually fish were not a favorite food of the Utes and were utilized mainly when there was little other meat. When the Utes did fish, they used lines made from the ligaments of elk and hooks of splintered bone. The Utes built basket traps of willow branches to funnel fish into a pool where they could be netted. Fish were usually not eaten immediately, but were split down the middle to the backbone and then laid across two poles to dry. Dried fish were stored in skin sacks and placed in pits lined with bark or grass. The pits were covered with branches and bark and then covered with rocks and dirt to keep the meat cool, dry, and safe from predators. Most surplus food was stored this way.

Rabbits and antelope were a major source of food for the Utes. Men hunted rabbits with a bow and arrow, a sling, a throwing club, or with snares made from sinew. To kill antelope the Utes built wing fences of rock, tree limbs, and brush. Hunters then drove antelope along these fences to ambushes where men used arrows, spears, or clubs to kill their quarry.

The Utes hunted buffalo on the Colorado prairies and in the high mountain parks, including the San Luis Valley. When the Utes killed a buffalo, it was like one-stop-shopping, as every part of the animal was used to provide either shelter, food, clothing, implements, or other necessities of life. Not a fiber was wasted.

Bighorn sheep were surrounded and driven up a mountain to hunters who were waiting in ambush. When a group hunted together, the person who killed an animal was entitled to the hide, but the meat was divided equally among the people.

In the winter deer were hunted close to their trails and ambushed from nearby cover. Elk were surrounded and driven toward deep snow until they tired and floundered, allowing hunters on primitive snowshoes to get close enough for a kill. Elk and deer meat were pulled to camp across the snow on top of the hides.

Predators such as the grizzly and black bears, wolves, foxes, coyotes, wolverines, mountain lions and bobcats were considered spiritual brothers. They were seldom hunted unless food was scarce. The lion was considered the bravest of all animals and the bear was second. It was the bear, however, who was the most powerful and was held in special spiritual reverence by the Utes.

The Utes lived in the high country seasonally as members of the ecosystem and just like other predators they experienced feast and famine. When game was not plentiful in an area or when winter snows moved the game, the Utes also moved. Wildlife, particularly big game populations, were regulated by severe winter conditions, disease, and predation more than by the Ute hunters. They took some and left some, and nature renewed itself in its own time and cycles. The seven bands of Utes that lived in western Colorado didn't need to change the lands or the waters that the Creator had provided them. The wealth of a man was measured by how much he shared with others. Ownership of land or its resources was a foreign concept. This was a big and bountiful land that sustained the Utes.

The Europeans: Trappers and Traders

The San Juans Mountains and the Ute Indians were formidable obstacles to European settlement and exploitation. However, the winds of change came to the San Juans in the early 1800s when it became the last mountain range in Colorado to be explored by Caucasians. Wildlife played an important role in the early exploration of the West. European fashion created a market for beaver top hats and by the 1820s the beaver had been trapped out of many eastern rivers, but were still abundant in the western rivers. Having successfully traded with the Spanish for horses and firearms as early as the mid-1600s, the Utes took the opportunity to trade their hides and furs with the trappers. By the early 1820s there were about 100 trappers working along the Rio Grande and almost as many on the Western Slope of Colorado. Some of the early trappers became famous, such as Kit Carson who signed up along with Jim Bridger to trap for John Jacob Astor's American Fur Company in 1834. Otto Mears, who became well known as the toll road and railroad builder of the San Juans, was also a beaver trapper.

In the late winter of 1821-22 Jacob Fowler, a trapper from Arkansas, led a trapping expedition up the Arkansas River. Fowler recorded his observations in his journal. As the party left the Arkansas and started up the Purgatory River, Lewis Dawson, one of the party, was attacked and killed by a grizzly near present-day Las Animas in eastern Colorado. The party made it into the San Luis Valley during the winter of 1822 and started trapping along the Rio Grande. The party caught beaver along the frozen bottom lands of the Rio Grande and set up their main camp below the cliff at Alpine just east of present day South Fork. From this camp they killed beaver, deer, elk, bear, geese, a sandhill crane, and a river otter. They looked for beaver signs up to what Fowler called "Hot Spring Crick" (which would be renamed Goose Creek) south of Wagon Wheel Gap. They were fascinated, even scalded, by the hot waters but they didn't find any beaver. They didn't explore any farther, because of deep snow. They did kill two elk between Wagon Wheel Gap and South Fork. Fowler noted that there were Spaniards and other trappers in the area. On March 7, 1822, they had a surprise waiting for them:

> . . . on our return we see a nomber of Ideans at camp which we cold see at some distance from the point of one of the mountains and not noing what Ideans they were we viewed them about half an hour-then we moved off to our camp and we came in . . . Tho the Ideans had stolen two buffelow roabs some lead and two knives. They were of the utaws (Ute) nation which roame about and live in the mountains without having any settled home and live alltogether on the chase . . .

Although Fowler did not record the presence of buffalo or antelope in the San Luis Valley, they were probably the most abundant of the big game species at that time. Antoine Leroux, a Taos trader, recorded that in 1820 there were thousands of buffalo and antelope in the San Luis Valley. Zebulon Pike's expedition into the San Luis Valley in 1806 killed eight buffalo to feed themselves. Buffalo bones have been found from valley floors to alpine areas in southwestern Colorado. Stuart Erickson, a Forest Ranger on the Alder District, found an ancient buffalo skull in 1928 on the south side of Pool Table Mountain. Buffalo were present on the Western Slope of Colorado from the present day Grand Junction area down into the Four Corners area. Their migrations beat out some of the trails between mountain parks that were used by most of the early explorers and prospectors. For example, the word "cochetopa" is the Ute word for "buffalo gate" which indicates that Cochetopa Pass, located to the west of Saguache, Colorado, was a buffalo trail between the San Luis Valley and the Gunnison country.

During this same time Antoine Roubidoux established the first trading post in Colorado below present-day Delta. Roubidoux facilitated the flow of furs to eastern markets by trading supplies with the Utes and Caucasian trappers until his trading post was destroyed by angry Utes in 1844. The fur market had crashed and the Utes did not understand why their furs had lost value. They were also angry about the invasion of white people into their land.

Yet there were trappers that the the Utes tolerated and even accepted. Bill Williams, whom the Ute Indians called "Lone Elk," had been a trapper in southwestern Colorado and was with the ill-fated 1848-49 Fremont expedition which attempted to find a route over the Continental Divide. Deep snow caught the party in the high country north of present-day Del Norte and above the habitat of any wintering game. Most of the party nearly starved to death and the survivors had to kill and eat some of their pack animals before they could retreat to lower country.

Prospectors and Wildlife

Between the collapse of the beaver trade in the 1830s and the discovery of silver and gold in the San Juans in the 1860s, there were few white people going into the San Juans. When precious metals were discovered, however, prospectors surged to the area by following the Rio Grande to its headwaters, crossing the Continental Divide at Stony Pass, and traveling on to the various mining districts. This was the major access route until the late 1870s. There was no nearby agriculture nor transportation system to provide food and supplies. Agricultural towns such as Montrose were not established until the 1880s. Railroad towns such as Durango and Gunnison began at about the same time. As a result, the prospectors lived off the land just as the Utes had been doing. But now the human population was far greater and modern firearms provided the means to kill wildlife like never before.

George Howard, a prospector after whom Howardsville was named, kept a detailed journal of his exploits in the San Juans. He left Cimarron, New Mexico, in April of 1872 with four burros, a pony, double barreled shotgun, Winchester rifle, six-shooter, and six months of supplies and headed to the San Juan mining district. Excerpts from his journal illustrate the importance of wildlife to the prospector. Passing through present-day South Fork he killed a "prairie chicken." On May 6, 1872, he left South Fork and traveled up the Rio Grande and wrote: "I saw several small bands of antelope today, but did not get a shot."

Later he moved up river to his next camp:

> My camp is a beautiful one . . . on the bank of the river in a fine little park surrounded on all sides by high mountains . . . grand and majestic . . . This place is called by some the "Wagon Wheel Gap."
>
> Goose eggs are no bad thing to take, so I found today a dinner. The two old geese flew by camp in the evening. I turned the old shot gun loose at them, and broke the wing of one of them, but it fell in the river and escaped me.

Howard most likely named Goose Creek that flows into the Rio Grande at Wagon Wheel Gap.

In a few days he made his way to the confluence of Ute Creek, Lost Trail Creek, and the Rio Grande where he as well as many other prospectors waited for the winter snowpack to melt enough that they could cross the Continental Divide. While they waited, they fished for trout and shot ducks and geese on the beaver ponds. They also hunted bighorn sheep on Pole Mountain.

On May 21, 1872, Howard rode back to Antelope Park to try to shoot an antelope to increase his meat supply. He saw three, but didn't get a shot. After a week he had given up his antelope hunt and made his way upriver to Pole Creek. There he wrote:

> Several mountain sheep were shot last evening by some of the other parties, so as we were not going to move today, nearly everyone in camp went out sheep hunting, me included. Some of the others who were ahead of me, were more lucky and four or five sheep were bagged by them. A band passed within rifle shot of camp in the evening but nobody shot at them . . . I followed up some sheep this fore noon which passed in sight of camp and fired some 10 or 12 shots as they ran, but did not get any until I got them cornered up in a canon when I killed an old ram and could have got more but I run out of cartridges.

The next day he fleshed the sheep hide using the brains to tan the hide. The bighorn was a major food source for the early day prospectors. To preserve the meat they dried it for three or four days by hanging it on their lash ropes.

A summer of prospecting took Howard and four other men who joined him over into the San Miguel River drainages where they spent time hunting with little success. They did find the fishing in Trout Lake on the San Miguel very good. In one day the five of them caught 220 trout. They took the fish back to Cunningham Gulch and sold them for $1.50 a dozen.

After a summer of prospecting for gold and silver, Howard and his party of prospectors left for the winter on October 2, 1872, and followed the Rio Grande back to the San Luis Valley. Along the way Howard recorded that ". . . Mr. Kimble saw a large grizzly below Pole Creek, but was afraid to shoot at it, he being some distance ahead of us and alone."

It is interesting that Howard never recorded seeing elk or deer in the Upper Rio Grande or in the western San Juans where he prospected, yet others recorded seeing them. General William Palmer, owner of the Denver and Rio Grande Railroad, hired Henry M. Bennett to provide meat for the railroad builders in the San Luis Valley. In 1871 Bennett reported elk still being plentiful in Blue Park northeast of Wagon Wheel Gap. When elk were seen they were generally in small and widely scattered herds from the San Luis Valley floor to the head of the Rio Grande as reported by P.R. Bob Griswold in 1891:

> The elk roamed the forests and came down to the meadows just like they owned them, and they did. The beaver cut the trees and built their dams and homes along this scenic road to Silverton with colorful wild flowers along the way from Stony Pass to the Valley floor.

Charley Mason, a brother-in-law to John, Richard, and Clayton Wetherill (who all played a part in discovering the Indian ruins in what is now Mesa Verde National Park), came to the Upper Rio Grande in search of a location where he could develop a commercial fishery. He told Mabel Wright, who in 1916 was the first schoolteacher at Hermit Lakes, that when he first came to South Clear Creek to homestead the Hermit Lakes in about 1880, he saw many elk antlers and bones, but was puzzled that he saw no elk. From the 1820s to 1900 there were no records to indicate an abundance of elk anywhere in southwestern Colorado.

Bristol Head Mountain at 12,570 feet towers above Antelope park where nearly a thousand antelope ranged until about 1883. By 1885 prospectors had killed all of them. Sketch by Mike Simpson

The early explorers noted primarily those animals which were considered "game" and were used for food, hide, or fur. Naturalists such as H.H. Henshaw, who accompanied the Wheeler Expedition of 1873 and 1874, made extensive surveys and recorded the scientific names of those species that they could identify. Henshaw's 1875 report was the most comprehensive ornithological investigation conducted in the San Luis Valley. C.E. Aiken, Henshaw's assistant, added 104 bird species to the checklist for just the San Luis Valley. This inventory of wildlife species was an important accomplishment. It was particularly interesting that Aiken noted the presence of Merriam's turkeys in the foothills on the west side of the San Luis Valley and as far up the Rio Grande as Wagon Wheel Gap. The area was not typical turkey habitat compared with the San Juan Basin and Uncompahgre Plateau which are historic turkey habitats in southwestern Colorado.

Many antelope lived on the San Luis Valley floor and there was one large herd in the Upper Rio Grande. Jim, Dan, and Frank Soward homesteaded sections of Antelope Park in about 1876. Emma McCrone said that her uncle Jim had told her, "When I homesteaded in Antelope Park it was alive with antelope."

In 1878 Oliver North, a writer from England, reported in The Country Gentlemen's Newspaper, ". . . There is excellent shooting in the early fall on the Upper Grande, the hunting of antelope is superb at Antelope Park . . ." Henry Wason, who homesteaded the Wason Ranch in 1880, reported seeing more than a thousand antelope in Antelope Park as late as 1883. But by 1885 the last of the antelope had been killed by prospectors.

Treaties with the Utes

By the 1860s the Utes were becoming very disturbed by the increasing number of prospectors and settlers who were violating the terms of their existing treaty. Prospectors wanted to pursue the rich minerals that had been recently discovered in the San Juans, but they would be trespassing into Ute territory. The prospectors in the San Juans had good reason to be nervous, and early on they were usually not bothered. But the Utes began to visit the whites more frequently and were blamed for setting fires, stealing horses, and other alleged atrocities. The press, utilizing mostly fabricated stories, fanned public sentiment against the Utes. In 1868 the federal government signed another treaty in which the Utes retained much of western Colorado between the Utah-Colorado border and the 107th Principal Meridian-which is a north-south line that passes northward from Pagosa Springs through Creede, Gunnison, and Glenwood Springs to the White River. Unfortunately prospectors and settlers continued to trespass on Ute land and did not comply with the terms of this treaty.

In 1874 Felix Brunot, President of the U.S. Board of Indian Commissioners, negotiated another Ute treaty which was named after him. Chief Ouray tried to save all the land and resources that he could for his people. The Utes knew of atrocities that white men had committed against other Indian tribes, and it was becoming obvious that they could not withstand the overwhelming tide of white people closing in on the western Colorado Territory, and particularly the San Juan Mountains, which threatened to destroy their traditional way of life. Ouray agreed to concessions, decreasing the reservation from 23,000 square miles to 5,300 square miles. The cession permitted mining for gold and silver near the present-day towns of Ouray, Silverton, Lake City, and Durango. In the Rio Grande drainage the Brunot Treaty area included the lands west of Creede. The 1874 Brunot Treaty allowed

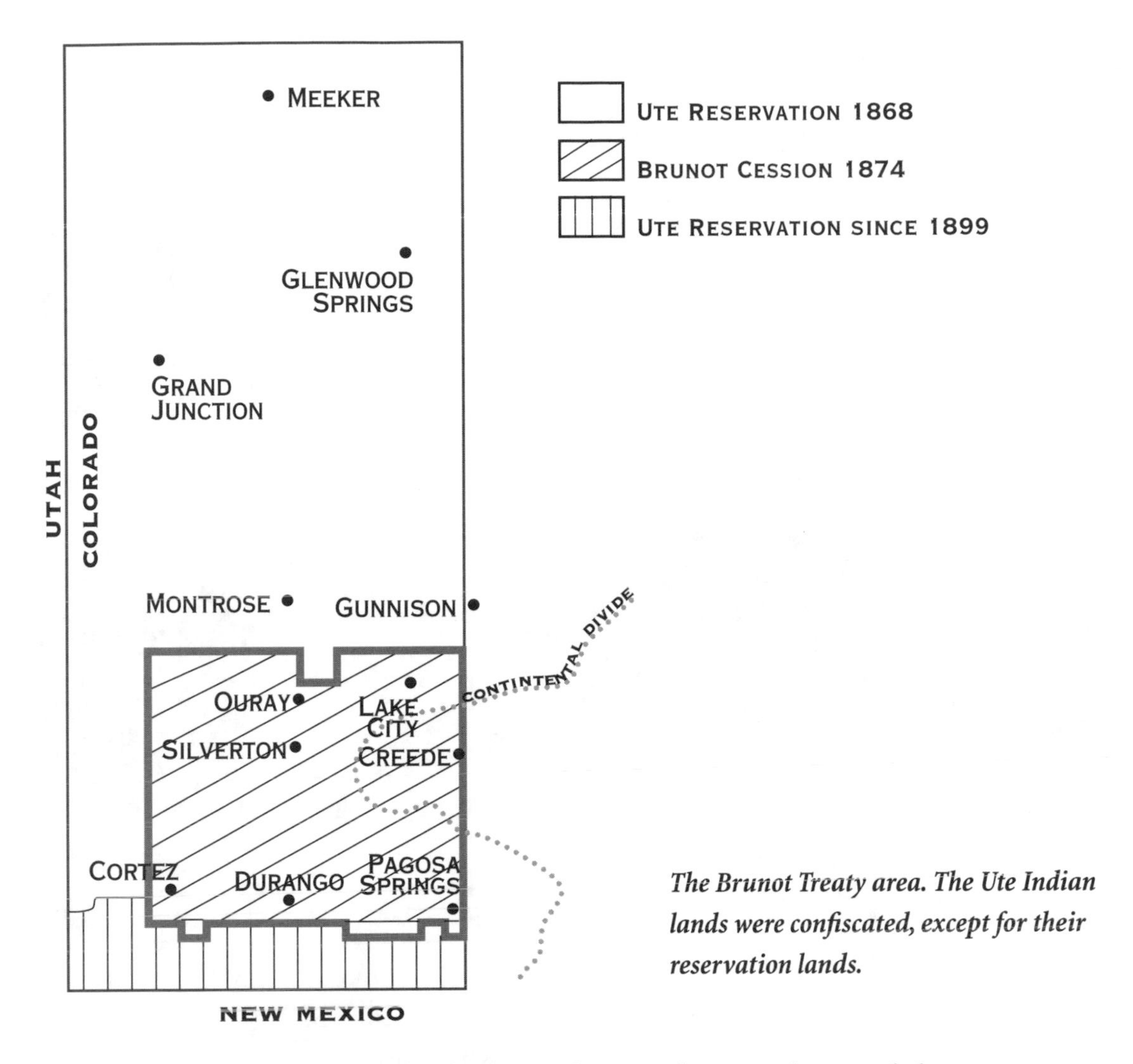

The Brunot Treaty area. The Ute Indian lands were confiscated, except for their reservation lands.

the Utes to hunt upon said lands, ". . . so long as the game lasts and the Indians are at peace with the white people." This issue would surface again a hundred years later.

In 1881 the last of the Utes were forced to reservations. The Weminuche band was forced to a reservation where they became known as the Ute Mountain Ute Tribe with headquarters in Towaoc south of Cortez, Colorado. The Mouache and Copote bands were moved to the Southern Ute Reservation headquartered in Ignacio, Colorado. The Tabeguache band was moved to present-day southeastern Utah. The Northern Ute bands that lived in the Yampa River drainage were moved to Fort Duchesne, Utah. The Utes actually needed to move since after 1880 the game which they hunted had been destroyed to the point that it was difficult for them to find enough for subsistence. Agriculture was not part of Ute lifestyle or

A fishing party enjoying the Rio Grande at Wagon Wheel Gap in 1883. Fishermen traveled by train to General Palmer's hot springs resort on Goose Creek south of Wagon Wheel Gap. Photo by William Henry Jackson, courtesy of the Colorado Historical Society

particularly to their liking, but was forced upon them. The United States government confiscated the Ute land and agreed to furnish some food sources such as cattle and flour, but they had to continue to hunt, because they didn't have enough food.

To add insult to injury the white man blamed the Utes for the decline of wildlife as reported by the *San Juan Prospector* in 1889:

> . . . the band of Utes who have been hunting lately on the head of the South Fork left that section more recently with eighteen elk which had been killed. If this is true, it is safe to say that the noted elk band that has heretofore ranged near the head of the South Fork has received a terrible blow, if not wiped out entirely. Old hunters will learn of the Ute slaughter with much sorrow.

It is interesting that when the press reported a Ute hunting trip it was "slaughter," but white people's successful hunting trips were considered to be manly ventures to provide food for the table.

The Popularity of Fishing

In the late 1800s fishing gained popularity both for the sport and the food. People who traveled from the front range of Colorado or the San Luis Valley figured on catching enough fish to make the trip worthwhile. Some front range people were affluent and had leisure time. General Palmer extended the Denver and Rio Grande Railroad to Wagon Wheel Gap in 1880, so people would stay at the hot springs resort managed by J.C. McClellan. In 1885 the resort offered many experiences including:

> . . . the serving up all the deliciousness in royal style, including . . . the delicious Rio Grande trout, which are often landed from the river, flopped in the pan, done to a turn, and on the table, all within the hour.

San Luis Valley residents brought their wagons and horses and made practical use of the fisheries, mixing pleasure with bringing in fish for the table. In 1899 the *Center Post Dispatch* reported, "The Wills camping party returned from Antelope Park the first of the week after a stay of ten days. There were 30 in the party and they returned with 2,400 fish." Fish were preserved by salting them in kegs and by smoking. This practice was common all over Colorado.

Fishermen were seeking out choice fishing areas throughout the state and homesteading lakes and river parcels for their exclusive use. One group of Kansas fishermen sent George Cade to the Upper Rio Grande in 1891 for the specific purpose of homesteading a place where they could go trout fishing. Cade filed the original homestead which was located immediately downriver from the Soward Ranch. It was called the Kansas Club for many years.

Regardless of where anglers came from, Colorado's fisheries were being decimated by the sheer numbers of fishermen who killed fish as if there were an unlimited supply.

Devastation to Wildlife: Homesteading and Ranching

By the end of the nineteenth century the white man had made his impact on wildlife populations. Elk, bighorn sheep, grizzly, antelope, and deer were nearly gone. The buffalo was extirpated from Colorado since 1879. It was a sad commentary for San Juan geologist T.A. Rickard to write in 1902, ". . . the Utes are gone from the mountains and so is the game which they hunted; that too has been driven away by the restless prospector."

The belief that the early prospectors and their rifles were solely responsible for the decline or extinction of wildlife in southwestern Colorado is rather simplistic; however, the Caucasian population that settled the San Juans exploited all its natural resources. There were only three known elk herds left in southwestern Colorado by the early 1900s. One small elk herd of about twenty head remained on the Hermosa drainage north of Durango. The other two herds were found on the west side of the San Luis Valley as recorded in *The Journal of Mammology:*

> The once numerous elk of the San Luis Valley were deprived of this good range area when the land was taken for settlement. Most of the elk were killed, but small groups retreated to the upper reaches of Saguache Creek and the Rio Grande drainages . . .The Rio Grande herd found asylum in the Goose Creek area to the south of Wagon Wheel Gap. Estimated at less than 100 elk in 1907.

To encourage settlement in the western United States Congress passed The Homestead Act of 1862 which required development of the land and the extraction of resources from it. This act provided an individual the opportunity to own 160 acres of land by living on it for five years and making improvements such as building a house, fences, irrigation ditches, and a garden. Homesteads account for most of the ranches in Colorado. Most of these homesteads were located in bottom lands where ranchers grazed livestock, developed irrigation, and put up enough hay to feed their stock through the winter months. Ranchers also grazed adjoining public lands, because they needed summer range for their livestock so that they could use their meadow lands to raise hay. In such a cold and dry climate 160 acres was not a large enough parcel of land to sustain a family. They grazed public

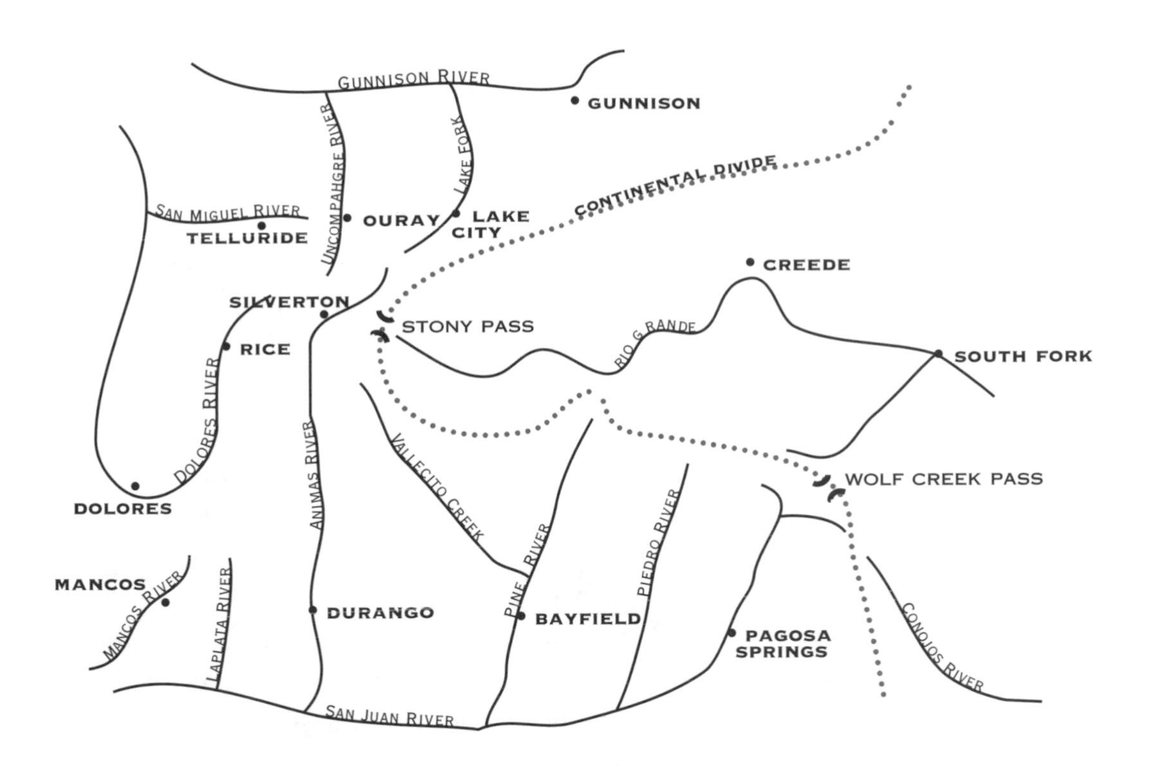

The San Juan Mountains are headwaters to hundreds of small streams that are tributary to the San Juan, Dolores, Gunnison, and Rio Grande Rivers. Nearly 150 miles of the Continental Divide surrounds the Upper Rio Grande like a horseshoe.

lands without restriction until the Congress established the national forests. Then ranchers applied to the Forest Service for grazing permits adjacent to their private lands.

Cattle and sheep operations were originally begun to meet the prospectors' and miners' need for meat. The earliest ranches were well established in southwestern Colorado by the late 1870s. Ranchers grazed as many cattle and sheep as they could. In 1903, when Congress was proposing the creation of the San Juan Forest Reserve, there was a field recommendation that the number of sheep be reduced by more than fifty percent. The sheep industry was lucrative at that time and there was a tremendous demand for sheep ranges. The cattle ranges were in better shape so the Forest Service recommended that the same number of cattle be maintained. The peak of livestock grazing occurred on most national forests during World War I and then again during the Second World War. Although the number of livestock on most national forests has decreased from these peak periods, most grazing adjustments have been in the form of shortened grazing seasons.

Frank Swank, who homesteaded Castle Lakes (which are presently called the Pearl Lakes) on North Clear Creek in the early 1900s, told his son Joel, "The sheep will overgraze this range and the bighorn sheep will starve out and disappear." Joel said that by the 1920s the bighorn sheep were gone from the cliffs on the west side of the Castle Lakes just as his father had predicted. Swank's prediction seemed to hold true for many bighorn populations in Colorado. Domestic and bighorn sheep compete for the same forage. Domestic sheep also transmit deadly Pasteurella pneumonia to which bighorns have never become immune. Bighorns are also an alternate host for a lung worm parasite that weakens them and makes them susceptible to pneumonia. Increased mortality from forage competition, disease, and overhunting killed off most of the bighorn sheep in southwestern Colorado, except for remnant populations on Pole Mountain and the La Garita Mountains. Other isolated populations struggled for their existence along the Continental Divide between the southern San Juans and Ouray.

Emma McCrone remembered seeing bighorn sheep in about 1900:

> When I was a girl, I remember herding cattle with my father over on Long Ridge and Seepage Creek and I saw as many as fifty bighorn sheep at a time. Father would kill one every so often for food. Even though we had cattle, we would vary our menu with wild game, especially in the summer when we had more ranch hands working for us. There was no regulation on them in those days.

Bald eagles used to soar high in the blue skies over the Upper Rio Grande. Emma McCrone said:

> They nested up on Bristol Head mountain. I also used to see them in the summer when we would ride in the buggy to Creede. We would see them down by the river getting a drink at the McCall place which is now Bristol Head Acres. They were so beautiful and I never got over how they could spread their wings and go higher and higher.

However, bald eagles soon began to disappear. For many years ranchers killed both bald and golden eagles, because they were

Emma McCrone was one of the pioneer women who could be comfortable entertaining a queen in the morning and stringing barbed wire in the afternoon (at the Soward Ranch circa 1956). Photo from Margaret Lamb Collection

thought to kill lambs. It was felt that one performed a community service by killing all birds of prey. Later the use of the pesticide DDT became responsible for the decline of several species of birds of prey.

Predation by mountain lions and bears on domestic livestock was a problem for ranchers throughout the West. Emma McCrone said that her father, Dan Soward, had problems with grizzlies killing cattle and young horses near the Soward Ranch in 1900:

> Yes, the grizzly killed our stock. The grizzlies were mostly back in the Trout Creek country between Middle Creek and Baldy Mountain. Before we had cows, father ran horses all through that country and the grizzly got after our colts, so he would get to the place where he knew the bear would come out and he'd set traps. He built a two-sided pen out of logs about four feet high and dug a hole for the meat and put the trap in front of the hole so the grizzly would step on the trap trying to get to the bait. A chain on the trap was attached to a log

The San Juans were the last refuge of the mighty grizzly in Colorado. The grizzly killed livestock and was extirpated from the Upper Rio Grande in 1951. Sketch by Mike Simpson

drag so he couldn't go anywhere. Father had two of those big grizzly traps. Well, old mister grizzly would come in to get the meat and he steps in the trap. I know all this, because when I was ten years old it was my job to check the grizzly traps, because the men were busy in the field putting up hay. After I helped mother finish washing the dinner dishes, I would ride my beautiful little horse up in the Love Lake country on Middle Creek. It was a great thing, almost a holiday to get on my horse and get away from my chores. I wasn't particularly afraid, because I rode a lot with my father. One thing I learned was that when there was a bear in the trap sometimes I couldn't get my horse within a quarter of a mile of the trap. Other times we'd come up over a rise and the bear would have the area around the trap all torn up, fighting that trap to get away. I felt so sorry for the poor bear. Anyway, I would get the fastest ride back to the ranch. I would tell the men who were working in the field and they'd get the wagon and went up there around Love Lake, kill the bear, and bring him down to the ranch. I've seen father and the men bring them down, skin the hide, stretch it, and nail it to the back of the barn. One would about cover the back of our barn. Those bears weighed 700 to 800 pounds. We would catch two or three a season. The last time I remember seeing a grizzly was in about 1910. I saw a grizzly in McDonald Park between Red Mountain Ranch and Spar City. He walked right through our cattle and they just watched him. They must have known he wasn't hungry.

Sarah Lamb, a ten year old great-granddaughter of Emma McCrone with one of the large grizzly bear traps her great-grandmother used to catch grizzly bears. The barn behind her is where the grizzly hides were stretched to dry. Photo by author

With a faraway look Emma said with thoughtfulness and sadness, "The grizzly . . . he kind of got lost didn't he?"

Any predator that was a real or perceived threat to the safety of persons or livestock was a target for extermination. Ranchers, especially woolgrowers, had great political influence on government policy and were able to get predator extermination programs started. Forest Rangers were directed to exterminate all predators, including blue darter hawks (goshawks) that preyed on grouse and other birds. Glen Munsell, a Forest Ranger between 1908 and 1911, recorded in his diary his attempts to kill coyotes. The Forest Service issued strychnine poison for coyote extermination. Munsell found that it was difficult to get just the right dosage of poison to kill coyotes.

The reclusive wolverine was practically extirpated from the San Juan Mountains by the early 1900s. A.C. Rowell, a trapper, said that he had caught a considerable number of wolverines in the headwaters of the Gunnison and Rio Grande Rivers and that he saw some wolverine tracks on the headwaters of the Rio Grande in 1903. Mr. Wood Galloway's father shot a wolverine in Antelope Park in about 1900 and said that they were common in the area. Since those early days, there have been only unconfirmed wolverine sightings in the San Juans.

Wolves

Wolves were seldom heard or seen by the early settlers in the Upper Rio Grande, but they were common in all other basins in southwestern Colorado. Undoubtedly wolves crossed between the San Juan and Gunnison basins, but there is no recorded witness to wolf packs roaming the Upper Rio Grande. There is a record of a wolf being killed in 1926 in the Creede area, but there was no mention of wolves in the Creede newspapers. Ranchers had no problems with wolves. Charles Dabney said:

> When my dad homesteaded on the E.J. Dabney Ranch on North Clear Creek in 1904 there were grizzly that killed his cattle. There were some black bears. I don't think there were ever any wolves in this country.

Howard Kennell, who was born and raised in his early years in Creede and became a state trapper and later a State Fur Inspector, reported:

> I never heard of a wolf in the Upper Rio Grande. The last wolf report that I remember was when I was the fur inspector stationed in Monte Vista. John Crooks, the Federal Fish and Wildlife supervisor, told me in about 1938 they killed a wolf over on the Navajo drainage south of Pagosa Springs and he claimed that was the last wolf in Colorado.

The woolgrowers continued to put political pressure on the U.S. Biological Survey (Fish and Wildlife Service) and Forest Service during the 1920s to eliminate predators. The extermination of wolves and trapping of mountain lions, bears, and coyotes in the San Juan Basin as well as other good ranges in southwestern Colorado undoubtedly helped deer and elk recover more quickly from the brink of extinction.

Devastation: Logging and Mining

Art Davis in about 1950. Davis grew up in Creede and was witness to 0the r00ecovery of much of the wildlife. Photo from Gordon Duncan Collection, courtesy of John Jackson

Creede was the last silver boom town in Colorado. The discovery of silver in Creede in 1892 brought more than 10,000 people into the Upper Rio Grande. This sudden increase in the number of people multiplied the demands for construction materials, just as it had done in every other mining camp in Colorado. In all those camps, loggers brought in saw mills and started cutting local forests that were milled into lumber to build houses, bridges, railroad ties, mining timbers, and used for other structures. Forests that were next to a town or to the mines were cut first. Entire mountainsides of spruce and fir were clear cut. Logging in those early days was done with axes and crosscut saws. Horses dragged the logs to waiting wagons which hauled them short distances to a nearby sawmill. The lumber was then hauled to town or to the mines. The back country was not exploited because the more easily accessible timber met the local needs.

Art Davis, a long-time Creede resident, remembered the loggers:

> My stepdad was a "tie hack" in the Deep Creek country south of Creede. They cut Douglas Fir trees and sawed them into lengths for railroad ties and skidded them down to the Deep Creek bridge on the Rio Grande. When high water came in the spring, the ties were floated down the river to Del Norte.
>
> The tie hacks cut a lot of the timber in that area and it was them who accidentally started the fire that burned through Deep Creek in 1890. That fire burned its way from Deep Creek and jumped the Rio Grande where the Wason Ranch is and burned across to Farmers and Bellows Creek. It just burned itself out, because they couldn't fight fire then.

Forest fires, whether started by man or by lightning, have been responsible for creating the diverse mosaic of forest and meadow in the San Juan Mountains. Most fires have been small, but some such as the

Lime Creek burn south of Silverton burned across thousands of acres.The sudden population growth placed greater demands not only on the land for timber, but also on wildlife for food and on range land for horses, mules, cattle and sheep. These demands strengthened the justification for making land use decisions based strictly on the economics with little consideration for wildlife. Mining laws gave a free rein to the mining industry. People took what they wanted from the land without thought of subsequent damage to land, water, and wildlife.

Just like the earliest mining operations throughout Colorado, later mines and mills dumped their acidic water loaded with heavy metals, mill tailings, and raw sewage into the nearest creek. Any local creek became a sewage "ditch" to the nearest river. There was an "Out of sight-out of mind" mentality. The milling processes loaded the local streams with very fine silt. Cutthroat trout did not survive these conditions, because they have very fine gill filaments that make them intolerant to long periods in muddy water. As a result, the cutthroat trout soon disappeared from those sections of water that were downstream from mining and milling operations. What a change from Henry Wason's report in 1883 as he stood on the banks of Willow Creek near Creede and counted more than a thousand cutthroat trout. He said that number was common for the mountain streams of the area at that time.

Killing wildlife for market was not as common in Creede, for example, as it was in earlier mining camps in Colorado, because the elk, deer, bighorn, and antelope had already been killed off by the time Creede was founded in 1890. By then agriculture and food distribution were well established in Colorado, and as a result there was less dependency on wildlife for food. Still people were opportunists and varied their menu with wild game and fish whenever they could.

By 1900 the prospectors, miners, ranchers, and general population had all but destroyed the natural environment of the San Juans. They did not set out to destroy nature, but their pursuits to extract their fortunes from the land had devastating consequences for the wildlife and its habitat. In the decades to come there would be further destruction and loss of wildlife, but attitudes were beginning to change. There began to be a desire to recover the wildlife heritage that had been lost.

CHAPTER 2

THE AWAKENING

Some knowledgeable people recognized that America's wildlife was in peril as early as the 1870s. Congress as well as state legislatures and private organizations acknowledged the dilemma and created agencies and programs to reverse the decline of wildlife. The issue that concerned most people was the decline of America's fisheries. Over-fishing, pollution, habitat destruction, siltation, mining, and other developments were destroying what were once productive fisheries. In 1871 the United States Congress addressed the issue of diminishing fish populations that were used for food by creating the position of "Commissioner of Fish and Fisheries." His mission was to increase the number of food-fish in America. The creation of this office would soon be mirrored by most state legislatures, and these offices became the forerunners of state wildlife agencies. Although recovery of fisheries was at the top of the national and state agendas, terrestrial wildlife was also on the decline and state legislatures, sportsmen, businesses and other citizens began to take action locally to protect remaining wildlife with new laws and projects to stop the downward spiral to extinction and bring back the wildlife.

EARLY HUNTING AND FISHING LEGISLATION

Colorado was granted statehood in 1876. In its first session the legislature of 1877 passed statutes that outlawed commercial hunting and limited the killing of game to personal food consumption only. They made no provision for law enforcement. There was also a grasshopper infestation at that time and agricultural interests asked the legislature to protect insect-eating birds. However predator control was very popular, and a bounty of twenty-five cents was given for killing each hawk and fifty cents for killing a wolf or coyote. These bounties were paid out of the state general fund.

The same Colorado Legislature created a State Fish Commissioner position. William Sisty was the first to hold this office and is considered to be the "Father of Fish Culture in Colorado." His mission was the same as his federal counterpart-fish propagation. Sisty's biggest claim to fame was the introduction of the rainbow trout into some of Colorado's waters in 1882. Because of its remoteness, southwestern Colorado's lakes and streams were among the last waters to receive rainbow trout. There is no clear record of exactly when the rainbow trout was first stocked into the major rivers of southwestern Colorado, but a seven pound rainbow was caught in the Conejos River in 1900, so it is apparent that rainbow trout were stocked in that river as fry or fingerling several years earlier. It is also unclear if rainbows were first stocked by private or federal hatcheries, but they were likely stocked as early as 1890 as reported in Field and Farm in that year:

> State Fish Commissioner Gordon Land, through his deputy, F.A. Ingraham, last week took to Del Norte over 2,000 eastern brook trout and California rainbow trout, and deposited the same in the headwaters of San Francisco Creek. It was the intention to place these fish into the San Francisco Lakes, but the stormy weather at that altitude prevented reaching the lakes with the trout.

These fish may have been received by early fish culturist L.D. Mercer of Del Norte. In 1891 R.H. Duckett of Gunnison sold his operation and *Field and Farm* wrote, ". . . Mr. Duckett expects to relocate at Del Norte and engage in fish culture with L. D. Mercer of that place." These two fish culturists were involved in trout stocking in the Upper Rio Grande. Rainbow trout were stocked about this same time in the Gunnison River as well as other major streams in southwestern Colorado.

In 1893 the Legislature passed statutes to protect all buffalo and bighorn sheep. The buffalo had already been totally extirpated from southwestern Colorado and there were only small herds of bighorns struggling to survive. Deer, elk, and antelope seasons were closed from August 1 to November 1. There were no bag limits, licenses, or any other restrictions on other species.

By 1884 the U.S. Biological Survey (precursor to the U.S. Fish and Wildlife Service) had built thirteen fish hatching stations across the country. Colorado's first federal fish hatchery was built in Leadville in 1889 and provided many of the the fish which were stocked in the state before state hatcheries began stocking fish in the Rio Grande drainage. The Leadville unit raised cutthroat, brook, rainbow, and brown trout. Even arctic grayling were stocked in the Rio Grande in 1899. There were no fish trucks or highways to facilitate the distribution of fish, but the U.S. Fish Commission built special aquarium-railroad-cars that distributed fish to streams and lakes along railroad routes.

In 1891 the U.S. Congress created the Forest Reserve System, which became the U.S. Forest Service in 1904. Because so many eastern forests had been exploited, there was a growing concern about the exploitation of western forests. The San Juan Forest Reserve was established in 1904 and included both sides of the San Juans as well as the La Garita and La Plata mountain ranges. The Forest Service would play an important land management role in the San Juan Mountains.

The Colorado Legislature became concerned about continued exploitation of wildlife resources and added to the duties of the Fish Commissioner. In 1891 the legislature gave him the authority to appoint five chief game and fish wardens who would draw a salary of $900 per year, ten temporary deputy game wardens at $100 per month, and special game wardens who served without pay. These officers patrolled the entire state of Colorado. F.B Orman, one of the first chief game wardens lived in Pueblo and patrolled thousands of square miles of southwestern Colorado on horseback in a country that had few roads and few towns, but many ranches, mines, logging camps, fishermen, trappers and some hunters. It was an insurmountable task. The legislature passed a few laws aimed primarily at protecting fish. Fish could be taken only by hook and line, fishing season was closed December through April, fish could not be wasted nor transported out of state, and it was illegal to sell fish taken in Colorado waters except for private commercial operations.

In 1897 the Colorado Legislature created the Division of Forestry, Game and Fish and the head of the division was titled "Forest, Game and Fish Commissioner." In 1899 the legislature deleted "Forestry" and changed the name to Department of Game and Fish, which lasted until 1963. The mission of that early department was specified as fish culture and wildlife law enforcement. The department was financed from the General Tax Fund, because fishing and hunting licenses had not yet been created.

The decline of the native cutthroat trout was blamed on commercial marketing of fish, dynamiting, fishing traps, nets and overfishing. Irrigation ditches were also a cause for the decline. They accounted for the loss of thousands of fish every year in Colorado rivers. Thousands of cutthroat migrated down these rivers and followed diversion ditches and out onto fields. For example *Sports Afield* reported in 1889:

> . . . The friends of trout and trouting must try to get the San Luis Valley farmers to agree upon some reasonable and practical policy, instead of antagonizing them by attempting to coerce them through laws which they are determined to regard as distasteful and tyrannical. It is for the farmers and men in authority to say whether the Rio Grande shall remain a trout stream or not . . . the hostile attitude of the valley farmers below Del Norte seems likely to defeat the establishment of a branch hatchery at or near Wagon Wheel Gap. While the ranchmen along the upper waters are ready and anxious to protect the trout, that will avail nothing unless the lower valley men will cooperate with them. The trout must descend into the deep water in the winter. If there is no deep water left for them to descend to, or if they are to be trapped if they get below Del Norte, there will soon be no native fish, and there would be no object in establishing a hatchery for imported ones.

The legislature was greatly influenced by agriculture and would not pass legislation that would inconvenience or counter agricultural interests. It is doubtful that screening could have saved the native cutthroats, but such losses contributed to the extirpation of this species from major rivers.

In 1897 the legislature established the first bag limits, allowing a maximum (but very generous limit) of game and fish that could be possessed: fifty ducks (plus twenty-five other birds), twenty pounds of trout, and fifty pounds of other fish per day; trout had to be eight inches in length; and there was an open season on bull elk, buck deer, and antelope.

Though livestock producers were having their problems with predators, it was the feeling of wildlife professionals that the popular clamor to have bounties on lions, bears, wolves, and coyotes would not help the other wildlife. Gordon Land, Fish Commissioner, wrote in his 1889 annual report to Governor John L. Routt a rebuttal to the demands to exterminate predators:

> I would like to ask that no bounty laws be passed that embraces a premium on either bear or mountain lions. The claim made that these animals are particularly destructive to both stock and game is not well founded, since in the case of game, as is well known, the deer, antelope, and elk existed in far greater abundance when there were no such laws than they do now; and as for the killing of stock by mountain lions, I am of the opinion that the worthless beings, I will not call them men, who pursue and hunt these animals with dogs solely for the reward, are far more destructive to the live stock interests than are these solitary beasts of prey.

The livestock industry would prevail, and the legislature would eventually establish bounties for coyotes, lions, wolves, hawks, and other predators, which were thought to be harmful to agricultural interests.

Although the early laws protected wild fish and game from commercial sale, the legislature specifically allowed for private fish culture and gave that industry some protection of law. But as Emma McCrone said, her father's way to protect his investment at the Soward Ranch fishery was "a shotgun and a 'crusty' attitude."

Before the Game and Fish Department built its first fish hatchery in western Colorado, the private and federal fish hatcheries had already introduced non-native fish into all major drainages in Colorado. Prior to these exotic introductions the three species of cutthroat trout were the only native trout in Colorado. Colorado River cutthroat was in all of the drainages on Colorado's Western Slope. The Greenback cutthroat lived in the front range drainages on the Eastern Slope, excluding the San Luis Valley. The Rio Grande cutthroat lived in the Rio Grande drainage in Colorado and New Mexico. Prior to exploitation, pollution, introduction of exotic species and overfishing the cutthroat in each river basin lived in balance with other native species. Nearly every tributary provided unspoiled spawning beds and nursery habitat for tiny fingerlings. Lush meadows and stream-side vegetation kept the fragile mountain soils from washing into the streams, keeping the spawning beds silt free and the sheltering pools deep enough to protect the fish through the severe winter months.

The Beginnings of Wildlife Protection

About 1900 Americans began to turn from the exploitation of wildlife to the protection and recovery of its heritage nationally, statewide, and locally. Exploiters, polluters, and those whose economic interests were threatened by wildlife values were powerful voices who wanted to continue their destructive pursuits. Perpetuation of the environmental destruction was justified on the economic basis of protecting jobs and generating wealth. Opposition to exploitation was beginning to mount, but it would take until the end of the twentieth century for pro-wildlife interests to have enough political clout to overcome decades of land, water, and wildlife abuse.

Concerned sportsmen continued to put pressure on the legislature to pass more restrictive laws that would protect game and fish, and 1903 was a banner year for new Colorado wildlife legislation. "All game and fish . . . are hereby declared to be the property of the state . . . " This law is still a basic tenet specifying that the people own the wildlife and not any private individual. This declaration included wildlife, fish, and birds on private as well as public lands. The legislature amended the June 1 to October 31 fishing season to include the statement " . . . that the public shall have the right to fish in any stream in this state stocked at public expense . . . " It seemed only fair that if the public paid to have a stream stocked, then the public should have the right of access to those fish. This issue would rise again and again over the years.

The state legislature created the first hunting licenses in 1903 to fund the enforcement of the laws they were passing. This was the beginning of the Game Cash Fund that separated hunting and fishing license money from the General Tax Fund, which to this day does not finance wildlife management in Colorado. The first hunting licenses were also established: Nonresident general hunting was $25, nonresident one-day bird hunting $2, and resident hunting $1.

There was a fervor to increase big game populations, and there were proposals that would go as far as making the whole state into a game refuge. The tide turned from exploitation and destruction to renewal and protection. Big game animals also received attention from the Colorado Legislature. In 1903 the elk season was closed, and it remained so in most of Colorado until 1931. It was not reopened in the Upper Rio Grande until 1938. In 1907 the antelope and bighorn sheep seasons were closed. The only legal big game was buck deer during a very short season. The legislature even closed the deer season from 1913 until 1918.

The Colorado Legislature created the first nonresident fishing license in 1909 at a cost of two dollars. Fishing was free for Colorado residents until 1913. During this time there was a six- month fishing season that opened on Decoration Day. The first bag limits for fish started in 1909 and allowed an individual to catch twenty pounds of trout per day. Fines for violations were set and were to be distributed equally between the county treasurer, the game cash fund, and the person instituting the action-usually the game warden. This law remained in effect until 1937. The spoils system created a conflict of interest, but it was intended that a hungry game warden would get out there and catch poachers.

The United States Forest Service

There were so few game wardens that in 1905 the Regional Forester directed all forest rangers to cooperate with game wardens. They were to act as game wardens when appointed by the forest supervisor and the district game warden, inform all hunters and travelers of the local game laws and endeavor to prevent their violation (with courtesy and tact), and report violations to local game wardens.

Glen Munsell was District Forest Ranger in Creede from 1908 to 1911. As with most forest rangers in the San Juans his job responsibilities included scaling timber, writing timber sales, counting cattle on the public range, writing permits, taking care of horses, and riding long hours in all weather during all seasons. Of course even in those days there were reports to be written. He built and fixed phone lines. Munsell patrolled the country from Wagon Wheel Gap to Stony Pass on horseback. He was faithful to take Sundays off. He supervised the cattle and sheep ranges and told the ranchers to move their stock if it was overgrazing. He wrote tickets for livestock trespassing on the National Forest, but more often he just told the rancher to move the cattle or sheep and no summonses were issued unless the trespass was flagrant. He put out poison and traps for coyotes and bobcats. A forest ranger worked alone most of the year. From 1908 to 1911, Munsell's diary never mentioned deer, elk or bighorn sheep.

Most of the ranger's training was intended to enhance the development of the timber and grazing industries on National Forest lands. Munsell noted in his diary:

> *February 15, 1909* . . . read up on the *Green Book Stock Laws*. . . . *March 7* Wash day. Cut up enough firewood to last for a week. It takes all the wood a man can gather and cut in a day to last one week. . . . *March 15* Read three chapters of Fernow's *Economics of Forestry.* Can't say how much of it stuck to me. . . . *January 26, 1910* Read President Roosevelt's speech to the Forestry Congress held in Washington in 1905. Studied all about homesteading, fiscal proof and surviving on a homestead.

The Forest Service began training its forest rangers in the art and science of silviculture based on European traditions of extracting timber from the land. Forest rangers at that time were chosen for their horsemanship, integrity, resourcefulness, and independence, because they were isolated and had to make on the spot decisions without consultation with supervisors. They had guidelines and policies which were tempered by their common sense. Ranger districts were divided according to the size of the country and how much territory could be administered by a single ranger working on horseback.

In 1925 the Forest Service, U.S. Biological Survey, and local woolgrower associations entered into a cooperative agreement to intensify the campaign to exterminate all predatory animals. This agreement was continued for many years. The livestock industry pressured the legislature to require the Game and Fish Department to enter into that agreement in the 1930s.

Sportsmen Act to Improve Fishing

Soon after 1900 sportsmen in most western Colorado communities were organizing to improve the fishing and hunting. For example on February 26, 1910, the Rio Grande Fishermen's Association was formed at a meeting in Monte Vista, Colorado. Persons interested in improving the fishing on the Rio Grande and its tributaries were invited to attend. Representatives from all across the San Luis Valley attended. Theo. A. Wheeler, W.C. Duncan, and H.H. Wason traveled from Creede, joined by S.E. Land of the Phipps Ranch. Attending meetings in Monte Vista took a considerable commitment of time and effort considering it was winter and such a meeting would require at least two days, transportation being by train, wagon, or horseback. The Association committed to aid the authorities to protect the streams from all violations of the Fish and Game Laws, to endeavor to have screens placed in all the headgates to keep migrating fish from swimming down irrigation ditches to die in the fields, to hatch and buy additional fry to be placed in streams, and to make the Rio Grande and its tributaries noted for good fishing. One of the most aggressive efforts of the Association was to locate a new federal hatchery on the Upper Rio Grande. Such a project was considered a good thing for the entire San Luis Valley and southwestern Colorado in general.

By the second meeting on June 4, 1910, the Association had 250 members from all over the San Luis Valley. The Association had bought, and volunteers stocked, 100,000 Eastern Brook Trout in the Upper Rio Grande. Enthusiasm ran high as members believed they were making a difference and were definitely improving the fishing. A committee was appointed to pursue the building of a federal fish hatchery near Creede. Some citizens requisitioned fish and others took responsibility to get the fish stocked.

At this meeting the Association proposed the creation of a number of member fishing camps along the Rio Grande. Members applied to the Forest Service for the use of a small portion of land adjacent to the river for their exclusive use. Each camp was to be self-sustaining, and members would purchase their supplies at cost. These were to be base camps for people to live in, continue to stock fish, and otherwise improve the fishing. This was the beginning of what are now the summer home groups along the Rio Grande that remain to this day at Box Canyon, River Hill, and Lost Trail. By September of 1911 the Association was hauling fish to every stream and lake in the Upper Rio Grande as well as the entire San Luis Valley. The Creede Candle editor enthusiastically reported:

> The Rio Grande River Fishermen's Association received a consignment of 50,000 native trout fry from the Del Norte fish hatchery for distribution in the Rio Grande and its tributaries. A fish railcar of the

> Bureau of Fisheries in the charge of William Smith was attached to the train at Alamosa and was brought to Creede. It left several lots of fish along the way and brought up to us 37 cans of black-spotted cutthroat trout and four cans of rainbow trout fry for distribution. Agent Arthur Wason requested Mr. Wheeler, as representing the Rio Grande River Fishermen's Association to attend to the distribution.
>
> Several business houses donated the use of their teams and rigs, in fact there were too many, but their good intentions were much appreciated. Those that were used hustled the fish right out of town and put them in the places designated for them. Those in charge of the distribution used the greatest care in selecting places to deposit the young fish and all of them reported that the little fellows when left were having a great time. With all this fry placed in the river and creeks, conditions should soon begin to improve and better fishing should be available. Keep up the good work. Don't let it lag a bit boys!

Stocking fish was hard work and made for exhausting days. Ten gallon milk cans were used to transport the fish. Frequent stops had to be made to add fresh water from streams along the route. Glen Munsell, Forest Ranger, wrote in his daily diary of his days stocking these fish:

> *June 20, 1911.* Met fish car. Took seven cans from Creede to Farmers Union Reservoir. When I arrived at the reservoir, I found I could not ford the river to get to Squaw Creek with the pack animals. Was forced to plant the fish in the river above the dam site. . . . *June 21* Came from Reservoir to Creede to return the borrowed fish cans and buggy . . . *June 22* My team being fagged from their hard trip to the reservoir I stayed in Creede to let them rest up.

The "Fish Car Era" started in 1874 since there were no highways or motorized vehicles to haul fish any long distances. The fish railcars were built by the U.S Fish Commission (precursor to the Bureau of Sport Fisheries and the U.S. Fish and Wildlife Service) and were used for transporting fish and fish eggs from coast to coast. Each car could carry about 150 ten-gallon cans containing some 115,000 three-inch fish. A five-man crew attended to the fish and traveled in comfort. The water had to be aerated periodically and ice was used to maintain colder temperatures. These cars were used primarily to haul federally raised fish but periodically were used to transport state hatchery fish. The railroads provided fish delivery service free of charge. Recipients picked up cans of fish at the rail station nearest to where the fish were being stocked. If no rail terminus was nearby, a fish car messenger would unload twenty-five to thirty cans of fish and transport the shipment to a more convenient pick-up point. The applicants receiving the fish would be notified ahead of time by telegraph. Improved distribution with trucks on modern highways brought the fish car era to an end when the last of the fish cars was taken out of service in 1947.

Between 1918 and 1922 the state encouraged ranchers located along major rivers in Colorado to construct ponds that the state would then stock with fingerling trout. By the end of the summer fish had grown from five to eight inches and were then to be released into the river. On the Rio Grande such ponds were constructed at Wagon Wheel Gap, the La Garita Ranch, and the Freemon Ranch. Fish were stocked in all waters during this time, and some landowners began to lock the public out of stream sections which were being stocked at public expense. It took a few years for the public to become irate enough to take action, but in 1929 the Creede Candle reported:

> **Sportsmen Will War On State-Stocked Streams Not Opened For Public**
>
> Legislation to eliminate 'no trespassing' and 'no fishing' from Colorado streams will go south when sportsmen of the state meet in Denver this week.
>
> Trout streams would no longer be stocked at the state's expense and then closed to the public, if recommendations scheduled for endorsement of the Colorado Game and Fish Protective Association and the Izaak Walton League, meet approval in the state legislature.

Such legislation never passed, but later it became policy for the Game and Fish Department.

Anglers were concerned with water pollution in many of the streams in southwestern Colorado. In most early-day mining operations in Colorado the mines and mills dumped their tailings and toxic effluent directly into the streams. The Creede mines were no

The federal fish rail car system stocked fish in many rivers parallel to the railroad routes in the early 1900s. Here they delivered fish to cooperating citizens who then hauled the fish in milk cans to local streams and lakes. Photo from DC Booth National Historical Archive Collection

exception. There was an outcry from San Luis Valley sportsmen in 1910 concerning this problem. They demanded that action be taken to stop the pollution of the Rio Grande, but their concerns were ignored by the politicians. Many other rivers and streams in Colorado had already been polluted by mining operations and nothing was done to stop the destruction of the local fisheries. Laws protected the mining industry instead of paying attention to environmental concerns.

Reestablishing the Elk

Since there were few laws addressing wildlife problems, the perpetuation of fish and wildlife became a personal endeavor for some citizens. Reestablishing elk herds in Colorado from a few remnant bands was one of those endeavors. It took more than closing the hunting season to bring them back. Ranchers, businessmen, and sportsmen, worked with the Forest Service, National Park Service, Game and Fish Department and others to begin pushing efforts to bring the elk back into Colorado.

The Forest Service became involved in 1910 when Smith Riley, the first District Forester of the Rocky Mountain Region in Denver, sent a directive to Forest Supervisors:

> . . . to study the question of extending game to unstocked areas where there is every indication that certain species will thrive or where the infusion of new blood into the small bands, especially the elk, seems advisable. We must look forward to the time when the National Forests will become the recreation areas of this country.

Aldo Leopold, considered the Father of Modern Game Management in America, developed new concepts that became guidelines for the Forest Service regarding the management of wildlife on national forest lands. He proposed that the same scientific principles that were used to manage a forest be used to manage wildlife. Leopold reasoned that if one could measure the volume of timber in a forest, why couldn't the wildlife be inventoried and a sustainable harvest be calculated for hunting? He didn't propose how such inventories should be made, but the Forest Service attempted to comply with his recommendations. Forest Supervisors had their rangers begin gathering the first anecdotal records of wildlife conditions in the National Forests. Each Forest Ranger filed an *Annual Game Report.* Excerpts from these early reports describe the scarcity of wildlife in the San Juans. For example:

October 5, 1910, to W.J. Morrill, Supervisor, from F.E. Joy, Creede District Ranger:

> . . . the game in this particular section around Creede is rather scarce . . . most abundant game is grouse . . . deer: eight in Fir Creek, six between Rat and Miners Creeks, and four at the head of Dry Gulch and two between east and west Willow Creek . . . a few bighorn sheep and ptarmigan on Bristol Head Mountain and 14 bighorn at the head of Boulder Creek . . . Few elk at the head of Goose Creek and Leopard Creek.

Strangely missing in the *1911 Annual Game Report* was the effort to restock elk. The Upper Rio Grande was one of the earliest of such projects. Ranchers Dan Soward and Frank Coller, businessman Bert Hosselkus, and Forest Ranger Don LaFont, among others, took a lead role to restock elk in the Upper Rio Grande. Bertie Bruns, Creede pioneer, was an eye witness to the first transplant of elk into the Rio Grande,

Yellowstone National Park supplied the first elk transplant to the Upper Rio Grande in 1911. Elk were unloaded at Elk Creek and this load at the Wason Ranch. These elk mixed with the few remnant native elk to begin the recovery from near extinction. Sketch by Mike Simpson

The first time I ever saw an elk was when they unloaded them down at Wason Ranch below Creede. That would have been in 1911 the year before I was married. They had a pen there where they unloaded the elk. I saw them jump out of the railroad cars and they ran over across the river toward Deep Creek. I think there was about 35 or 40 head. They sort of disappeared and it was six or seven years after that we took a herd of cattle up to Lost Trail Creek and we saw some elk along the way. Whenever people saw an elk they always talked about it, because they were so new.

Roy Powell said that Don LaFont was the ranger when they released the first elk in 1911. Don went down to the Wason Ranch where he helped ear tag and release them. The same train left several railcar loads of elk at Elk Creek (just above South Fork) that day also. Frank Coller, who owned Elk Creek Ranch told Game Wardens Bill Schultz and Earl "Punk" Cochran that he was one of the instigators of getting the first elk placed on Elk Creek. According to Coller these first elk were unloaded at the Phillips Spur at the mouth of Elk Creek. Cowboys herded the elk up Elk Creek and within six months the elk were eating on his sheep range in Elk Creek Park.

There is no documentation in Forest Service or Division of Wildlife archives of this transplant having taken place in 1911, and the *Creede Candle* and other San Luis Valley newspapers never mentioned this occurrence. It has been generally accepted that the elk came from Yellowstone National Park in Wyoming. Yellowstone National Park elk shipments started in 1892, but there is no record of any shipment to Creede or South Fork. Congress authorized the sale or disposition of surplus elk and other animals in January of 1923. This suggests an informal approach to the process before then. Any work from 1911 to 1917 would have been done by the U.S. Army, who administered the Yellowstone National Park from 1886 to 1918, and they didn't keep records. In 1911 elk would have been baited with hay into corral traps during the winter and taken by horse-drawn wagon to the railhead at Gardiner, Montana at the Park's northern boundary. The railroad furnished free transportation, and in some other areas fraternal organizations such as the Elks Lodge, as well as the Forest Service, paid other expenses.. This scenario is very likely the one that occurred with the Rio Grande transplants.

The next year A.L. Sweitzer, Rio Grande Forest Supervisor, reported to the *Creede Candle* that in April, 1912, he and his staff made a six-day snowshoe trip into Elk Creek above South Fork to look for elk. He summed up his report:

> . . . That this is an ideal elk territory is clearly demonstrated by the reason that five years ago there was but fifteen or twenty head of elk in that whole country while now they have increased to at least 150 and also for the further reason that this is the only vicinity in the south part of Colorado where these animals are found to any extent.
> On account of the ideal conditions and the fact that the animals already are there, the government, the Elk Lodges and the public generally, should be sufficiently interested that they would all work together toward having set aside a game preserve in that territory for the protection and perpetuation of the species. This would not interfere with the grazing industry in the community and would add materially to the attraction of this part of the state.

The *1912 Annual Game Report* noted a behavioral difference between native elk and those brought in from Wyoming, which suggests that Wyoming elk were indeed brought in to the Upper Rio Grande:

> . . . native elk are in a different class from the Wyoming elk for the reason that it has never been necessary for them to come down to the

> ranches to winter and thus become accustomed to men on foot, horseback or in wagons. They are still wild as they were before they were interfered with by settlement and stock grazing. They will stay in the back country until their number increases to such an extent that they seek additional forage and come down to join their Wyoming cousins that have already discovered the feed on some ranches.

There may have been some morphological differences as well. Al Birdsey, a long time Creede resident said, "The first elk I ever saw was in 1922 on Goose Creek. He was one of the Colorado native elk. Those elk were much larger, darker color, and had much larger heads than the Wyoming elk."

In 1912 a transplant of elk was made north of Durango, Colorado, at the mouth of Hermosa Creek. The Forest Service estimated that only twenty head of elk were in the San Juan National Forest at that time. This elk transplant was well reported in a 1912 issue of the *Durango Herald.* In this instance the Forest Service, U.S. Biological Survey, and the local Elks Lodge split a bill for $625.53 to ship twenty-five elk from Jackson Hole, Wyoming. The Game and Fish Department apparently didn't participate in these earliest elk transplants since it only employed a few fish hatchery workers and five game wardens. The elk were shipped free of charge by the Denver and Rio Grande Railroad to Durango. The Wood and Morgan Furniture Store donated its moving van to haul the elk north to Hermosa where they were held for a month in special corrals and then herded by fifty horsemen up Hermosa Creek. These two transplants increased in number and range and eventually mixed with those few native elk that were left to become the base herd for the San Juan Mountains. There would be additional transplants in the 1920s in southwestern Colorado.

Winter—To Feed or Not to Feed

Even though elk were totally protected there has always been one limiting factor-the severe winters that play an important role in the status of all big game animals in Colorado. Winters in the Upper Rio Grande and Gunnison have always been especially hard on deer and elk. The Rio Grande is a grass winter range like no other range in Colorado. The Gunnison low country is mostly sage brush, except for the riparian areas where ranches exist. Winter ranges in the San Juan Basin and Uncompahgre Plateau, however, are among the best in Colorado, because of the diversity of browse species that deep snow does not bury. Elk and deer normally migrate to lower elevations to escape the deep snow, but in some Colorado winters there is no escape. After bottom lands and meadows were homesteaded, these historic deer and elk habitats were destroyed. Still the animals had to satisfy their nutritional needs. Soon increasing numbers of deer and elk began to have an impact on ranching operations throughout Colorado. For example in 1912 the *Creede Candle* reported:

> **Starving Deer . . .**
>
> The severe snow storms of late last week and early this week has created a condition among the game of this country not known in many years. Not less than one hundred deer have been driven down from their usual haunts in the Red Mountain country and around the head of the William's Fork into Antelope park. It is a most unusual thing for deer as wild as are these to approach anywhere near to where civilization exists, but the deep snows and the bitter weather that we have been experiencing of late have made the poor animals desperate . . . They were literally starving to death and it was this mad hunger which drove them down to the close vicinity of the Soward ranches and made them attack the huge hay stacks which were garnered there last fall. These stacks are protected by fences that are quite high, but the hungry deer easily clear them in their frantic efforts to get at something to eat . . . Mr. Soward, so we understand, strenuously objects to having his ranch used as a state game preserve and more especially since two of the deer broke into Mrs. Soward's milk house and licked up all of her dairy salt.
>
> There is one old buck, with at least eight points on his beautiful spread of horns, that is trampling the interior of the field and is standing off with savage attacks all who come nigh to him. Man or horse or both together, they all look alike to him and it being unlawful to use a gun on deer at this period of the year and there being no other means of securing him, he is a monarch of all he surveys and will probably remain so with immunity until the weather conditions are such that the herd goes back into the hills.

Another severe winter came in 1915, and the Forest Service began feeding elk in Elk Park by scattering hay along trails which had been broken through with horses. Many elk were forced to the river bottom between Wagon Wheel Gap and South Fork. Forest Service personnel said that as the number of elk increased it would be necessary at some future time for the government to provide winter feed for them other than the natural growth of forage on the range. The same thing was happening in the Gunnison country. Even in that day artificial feeding became the answer to solving the problem of nature's wrath. Early forest rangers and game wardens usually had an agricultural background, and their experience of caring for livestock naturally transferred to the protection of deer and elk. Little thought was given to the real problem of land development that had fragmented or destroyed the natural habitat that had sustained elk and deer and other big game animals for thousands of years. Feeding elk and deer in severe winters evolved into a way to mitigate the continued destruction of natural winter habitat. It artificially sustained them at higher population levels that satisfied popular citizen demand to have more animals for hunting.

Even in a fragmented environment the elk kept increasing. By 1919 the *Annual Game Report* reported that nearly 400 head of elk were in the area south of Wagon Wheel Gap. Even though elk were increasing, most people had still not seen one. Art Davis was a teenager in the early 1920s and had quite a surprise while hunting deer up Deep Creek south of Creede:

> I sneaked up on what I just knew was a big buck and I shot it and when I went over to it I didn't know what it was. I dressed it out and went down to our place on Sawmill Gulch south of Moonshine Mesa. I told my dad that I shot something and I didn't know what it was. Some men went up there with me to help pack it out and they stood around that big cow elk and argued over what it was. We packed it out, but the elk season was closed so we couldn't say anything about it.

During the 1920s elk continued to increase, but economic times were hard. Many mines were closed and employment was very limited. Soon elk poachers were at work between Elk Creek Ranch and

Goose Creek on the Rio Grande drainage. In one case Del Norte District Ranger Million was dispatched to investigate the elk poaching. Chief Deputy Game Warden Walter Campbell of Alamosa joined the investigation, but no one was ever caught. The Game and Fish Department wardens were criticized in the press for not doing their job of catching the poachers and they, in turn, challenged citizens to become involved in protecting elk:

> . . . The slaughter of these noble animals is in absolute violation of the game laws, which were so drawn that a closed season was declared for a number of years in order to give them a chance to propagate. Every good citizen, whether bearing a warden's commission or not should see to it that these killings cease immediately. The animals are quite scarce in these parts of the country and the count recently made by the Forest Service in connection with the Biological Survey showed that the herds in this vicinity had just begun to increase.
> The matter will bear the closest of investigation and if the reports are found to be true, the culprits should be punished with the utmost severity of the law. Anyone helping to catch them will be performing a public service.

Game Refuges

In the early 1920s Aldo Leopold started to promote the wildlife refuge idea as a panacea for protecting wildlife. The idea was to set aside small protected areas where wildlife would over-populate and the surplus animals would drift away from the refuge where they could be hunted. Little consideration was given then to protecting or managing wildlife habitat outside such designated areas. Refuges were created for waterfowl, deer, elk, and even fish.

As the elk increased through the 1920s, Leopold influenced the Forest Service to take the lead in setting areas aside for wildlife. Game refuges were established in several major deer and elk areas in western Colorado. These were soon abandoned, because the deer and elk expanded beyond these ranges and grazing interests wanted range for grazing. More refuges such as the Goose Creek Game Refuge were proposed, but died political deaths, because of pressure from agricultural interests.

More Elk Transplants

Elk transplants had been successful in the Durango and Creede areas and citizens in other regions also wanted to bring back the elk. Residents of Antonito requested that the Forest Service and Game and Fish Department transplant twenty-five head of elk to the Conejos River territory from the band on the Rio Grande. The proponents felt that the Rio Grande herd had increased sufficiently, and that it could stand to transplant some elk. The state was blamed for its lack of cooperation, and the transplant never took place.

While elk were recovering in the Upper Rio Grande, they were not faring so well in other areas of southwestern Colorado. During the 1920s elk were transplanted into several locations in Colorado. In 1923 the Game and Fish Department shipped in two railcar loads of elk from Jackson Hole, Wyoming, to southwest of Olathe, Colorado and released them to the Uncompahgre Plateau where the elk had been extirpated for many years. Another two railcar loads of elk were transported on over to Durango where they were released at the mouth of Hermosa Creek to supplement the transplant that had been made there in 1912.

Elk were transplanted from Routt County in northern Colorado to Ouray in the early 1920s and became quite the attraction for locals. Elk as well as bighorn often wandered the town in the winter and were jealously protected by the residents. When the winters were severe, local citizens fed the animals.

There was a second transplant of elk in the Upper Rio Grande during the 1920s. Frank Coller and the owner of the 4UR Ranch were the instigators of that plant on Elk Creek and Wagon Wheel Gap. The 4UR owned the hot springs on Goose Creek and had many of tourists who would ride up on the train to enjoy the resort. The resort wanted those people to see the elk also. Shorty Wheeler, long time resident, witnessed this second transplant:

> One day in about 1923 Dad came home and asked my mother and me if we wanted to take a ride from our place near Del Norte to above South Fork to see the elk that they just brought in from Wyoming. We went up there to Elk Creek Ranch and I saw five car loads of elk with about eight head each. They unloaded two car loads there and the others at Wagon Wheel Gap. The Game and Fish Department had them brought in.

Bighorn Sheep

Bighorn sheep were also getting some attention. Bighorn had been practically extirpated from many ranges in southwestern Colorado despite the fact that the hunting season had been closed since 1907. Edwin Bennett, Forest Ranger in the Upper Rio Grande, reported a remnant herd of about seventy-five head on Pole Mountain between 1919 and 1922. Ranger Elbert Vanaken reported in 1923 that the bighorn sheep were on the increase because the Pole Mountain range was off limits to domestic sheep and set aside for the bighorns. There were small groups of bighorns scattered in southwestern Colorado, but they were seldom seen by most residents. Alfred "Al" Birdsey, a Creede resident, got a sudden shock one day:

> I was hunting out of season in 1918 up in Shallow Creek with Charlie Lane. He had a 30-30 and I had an old rifle that Bill McCall's widow had given me. They had the McCall Ranch upriver where Bristolview Acres subdivision is now. We went out from Moonshine Mesa to Shallow Creek. By God Charlie shot a whole box of shells and never hit no deer. I was coming around the rimrock down by the sheep drive in Shallow Creek Basin and I had shot four or five times at deer and missed. Suddenly a couple animals jumped up and I threw that gun up and shot and damned if I didn't kill a mountain sheep ewe. The season was closed. Now that was an accident, but she tasted pretty good. The most sheep I ever seen in Shallow Creek was fourteen or fifteen head.
>
> Once I was riding horseback with Fay Franklin, our Forest Ranger, in the late 1930s and near the top of Bristol Head Mountain this old mountain sheep ran across in front of us with the biggest head I ever seen. I tried to get Fay to shoot it, because he said it wouldn't make it through the winter, but he wouldn't do it. That ram was coughing something awful.
>
> I think it was disease that got the bighorn more than hunting. They brought those domestic sheep in and the bighorns soon disappeared. John Dabney who was the caretaker at the Santa Maria Reservoir in the '30s told me that there were five big rams behind the main house that had died after being there just a few months and that was the last of them in that area.

Al Birdsey lived in Creede all his life, working, hunting and fishing and witnessing many of the events that affected wildlife. Here he displays a big buck taken in October, 1931. Photo from Dena M. Dority Collection

Bighorn sheep were also found in remnant herds north of Gunnison on the Taylor River and east of Lake City in the La Garita Mountains. There were small herds along the Continental Divide from the south San Juans to Ouray. Disease kept these herds from increasing until after the 1950s.

Wildlife management in the early 1900s was utilitarian and therefore most of the attention was being given to elk, deer, and bighorn sheep-those animals and fish that had economic value for sport and food. Protection from hunting and the reservation of some native habitat resulted in more animals being seen. Other species such as the black-footed ferret, river otter, mink, and grizzly, still faced extirpation. Maybe because of the isolation and rugged terrain, the San Juan Mountains held on to its wildlife longer than other regions in Colorado, but even the great grizzly would become a ghost. It would take several decades for some species to recover, and some never would because their natural habitat had been destroyed.

Chapter 3

THE PRIVATE FISH HATCHERIES

At the turn of the century fish culture in southwestern Colorado had become a major business. Although silver mining was on the decline, gold discoveries had picked up some of the slack. Agriculture and associated businesses brought economic and social stability and, as a result, there were more people living in southwestern Colorado. Transportation was improving and that increased tourism. One of the things people wanted to do was fish. When there was a need for food and recreation, there was money to be made, and there were those individuals who were willing to work hard to meet that need and make enough money to survive. There had been private and government operated fish hatcheries for many years prior to southwestern Colorado being settled. Therefore, it was inevitable that individuals would successfully learn the commercial fish business. In 1900 commercial fishermen were registered in Montezuma, Dolores, San Miguel, La Plata, Gunnison, Hinsdale, Mineral, Rio Grande, Saguache, and Montrose Counties. These individuals established an industry that would last more than a hundred years. Pioneer fish culturists further impacted on how Colorado's fisheries would be managed, because they were so successful at their trade. These early fish culturists worked closely with the new federal and state hatchery systems. Some of them became leaders in federal and state fish programs.

On just the Upper Rio Grande thirty-nine commercial fishermen produced fish in 1900. They reported catching more than sixty tons of fish for commercial sale. Some of these operations consisted of a hatching trough being placed in a spring where eggs were eyed (the stage where the eye of the embryo is visible) and hatched. Most operations sold eyed eggs or tiny fry, because they didn't have the means of transporting larger fish. Commercial fishermen obtained permits on public lakes and were allowed to seine fish for market. The Upper Rio Grande soon became one of the major commercial fish and fish egg producing regions in Colorado.

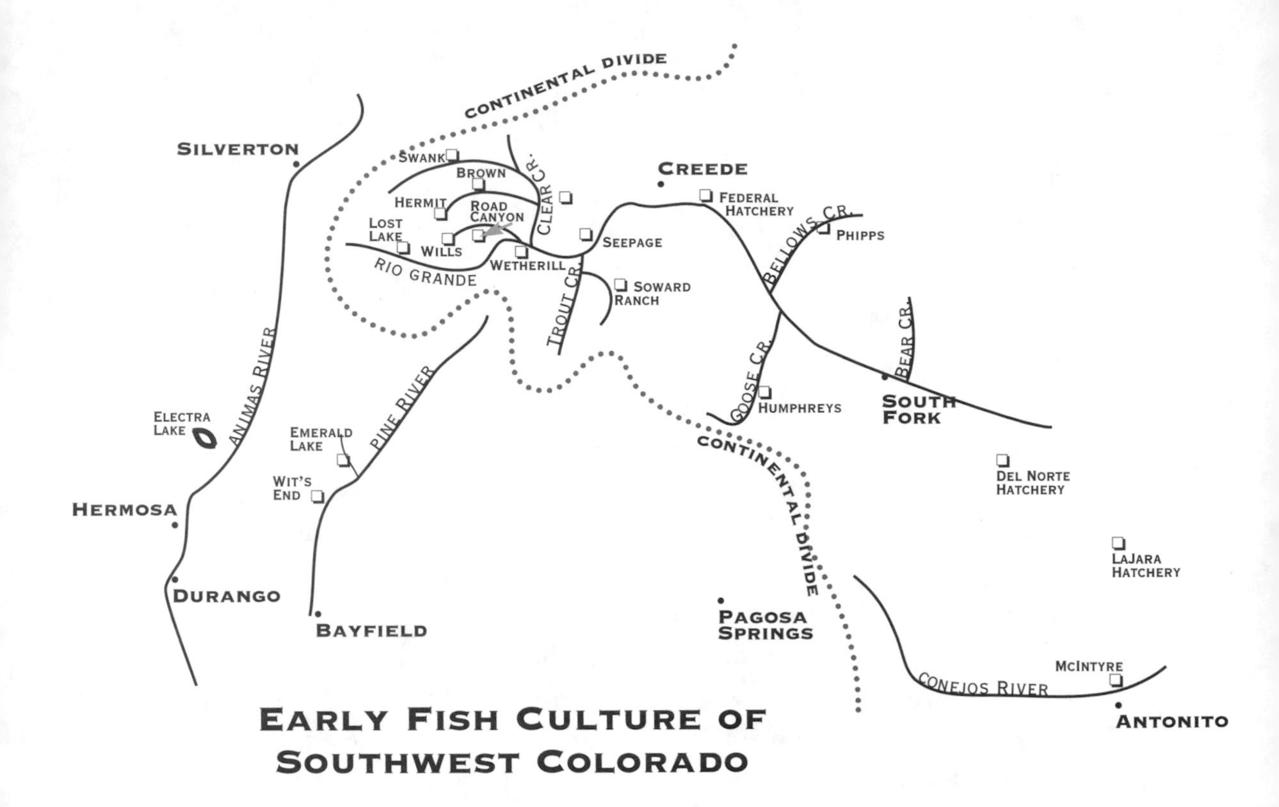

Pioneer commercial fisheries of southwestern Colorado were located at strategic locations where there was pure and warm spring water. All of these hatcheries raised trout. Map by author

Early Commercial Fish Hatcheries

Commercial fish operations in the San Luis Valley were started by Gordon Land in Conejos County in 1866. He became one of the leading fish culturists in Colorado and was appointed as the Colorado Fish Commissioner in 1889. Albert W. McIntire also owned and operated a fish hatchery and lake at McIntire Springs in Conejos County in 1880.

W.T. Kirkpatrick

Fish culture in southwestern Colorado lagged behind the earliest fish culture operations on the Eastern Slope of the state. In 1888 the Durango Rod and Gun Club seined Colorado River cutthroats from the Pine River and packed them on horseback in kegs up to the fishless Emerald Lakes on the Lake Fork of the Pine River. By 1895 W.T. Kirkpatrick, an early fish culturist from Durango, started propagating trout and by 1899 he was operating two hatcheries. Later, he allowed state spawning crews the use of his facilities at Emerald Lake, which resulted in the stocking of millions of Colorado River cutthroat trout into public waters throughout Colorado. Rainbow trout were introduced into Emerald Lake about 1900, and eggs were later taken by state crews until the late 1920s.

Patrick Brothers

Wash and Levi Patrick built the Wits End Hatchery and Trout-Rearing Facility on the Pine River in 1885. They raised rainbow trout in two small ponds utilizing spring water. They raised more than 50,000 trout annually, many of which were stocked in nearby lakes and streams. Wash Patrick went on to become superintendent of the state's La Plata Hatchery near Hermosa and in 1903 became superintendent when the new state Durango Hatchery opened. Between 1909 and 1913 Wash was General Superintendent for all state hatcheries.

Charles Mason

Charles Mason was the first fish culturist in the Upper Rio Grande. Mason and the Wetherills homesteaded ranches near Mancos, Colorado. They were cattlemen and known for their discovery of many of the cliff dwellings in what is now Mesa Verde National Park. Charles Mason and Richard Wetherill found the Cliff Palace while searching for stray cattle one day in December of 1888.

Mason associated with Wash and Levi Patrick at Wits End, north of Bayfield, Colorado. Undoubtedly the fish business appealed to Mason, and he spent time with the Patricks learning the business. According to pioneer Mabel Wright, Mason came to the Upper Rio Grande in about 1880 and maybe before that, but after he helped discover some of the Mesa Verde ruins, he had gotten a little tired of artifacts and ranching. He married Benjamin's Wetherill's daughter Anna. Mason began looking for a place where he could go into fish culture and build a lake or two without too much expense. He settled at Hermit Lakes on South Clear Creek up the Rio Grande from Creede.

Frederick Burrows had already homesteaded in South Clear Creek in 1862. Mason bought that property and added it to his own homestead in 1896. He bought another tract that included a natural lake. Later he purchased a second lake from Burrows and raised the dam of what is now lake number two. At that time one could take up a lake by homestead. In this way he enlarged Hermit Lakes to nearly 800 acres. Mason named the first lake that he built "Hermit," because he lived there alone. His homestead was the first to be patented in the United

Charles Mason in about 1920 with one of his sons. He was a pioneer fish culturist in southwestern Colorado. He built the first commercial fishery at Hermit Lakes on South Clear Creek, a tributary of the Rio Grande. Photo from Carol Ann Wetherill Getz Collection

States for the sole purpose of raising trout for market. He also had the first license issued for raising trout for market. His Number One license was retained throughout the years that his business continued.

Charles lived up at Hermit alone from spring to fall and returned to Mancos every winter. To cross the Continental Divide he rode the Ute trails from the Rio Grande over Weminuche Pass and down the Pine River. He moved his family from Mancos in the winter of 1902. After he built a fish hatchery he first tried stocking the native cutthroat trout, but it was not their nature to stay in the lakes. He shipped 100,000 eastern brook trout eggs from Plymouth, Massachusetts in 1901. The fish did well, so the next year he bought 200,000 and these brook trout became the source for stocking brook that we now enjoy in the Upper Rio Grande. Commercial fishermen began raising brook trout because the could take brook eggs in the fall and by spring the tiny fish were ready to stock. When those fingerlings were stocked in productive waters they would grow large enough by fall to be harvested for market. Mason stocked brook trout in upper meadows above North Clear Creek, because there were no fish up there. His operation was large enough that he built a school at Hermit. Mabel Steele Wright, who was raised on a ranch north of Lake City, was the first school teacher. Mason shipped trout and fish eggs to Denver and as far away as Australia.

Many of the private and commercial fisheries in Colorado were built on small streams with low dams so that the lakes could be drained and seined to recover the fish for market. Where possible, hatcheries were built next to springs that had warmer water than surface streams.

These early fish culturists were responsible for stocking many of the back country fisheries throughout southwestern Colorado. Carroll Wetherill reported:

Charlie Mason and my father Clayton Wetherill packed fish on horseback from Emerald Lake over in the Pine River country to the Rio Grande in the early 1900s. They were probably Colorado River cutthroats. There were no fish on North Clear Creek above the falls nor were there any fish in most of the high lakes in the San Juans such as the Ute, Rock, Flint, Vallecito Lake, Eldorado, or the Highland Mary Lakes.

John and Clayton Wetherill in about 1900. They were ranchers in Mancos, Colorado, and were explorers that discovered many of the Indian ruins in the southwest. Clayton, on the right, built a fish hatchery west of Creede. He and Charles Mason, his brother-in-law, were the first to stock fish in many of the high lakes in the San Juans. Photo from Carol Ann Wetherill Getz Collection

Frank Swank

Frank Swank homesteaded on North Clear Creek in 1909. He named his homestead Castle Lakes but it is now called Pearl Lakes. He built the dams with horse drawn scoops. Swank worked with Mason and Bert Hosselkus to develop fish culture in the Upper Rio Grande. Swank leased Black Mountain Lake from the Forest Service, stocked brook trout and then seined them to market the catch. Swank sold fresh fish to the Denver and Rio Grande railroad at Lake City and to various restaurants in town. Swank built another fish hatchery on Bear Creek north of South Fork to handle his surplus eggs. The Game and Fish Department purchased fish eggs from private hatcheries including Castle Lakes to hatch in its hatcheries. Later the state leased Swank's operation and took fifteen to twenty million trout eggs each year.

Hermit Lakes, in the background, was the first commercial fishery in the Upper Rio Grande. The two Brown Lakes, in the foreground, were built as a commercial fish operation and later purchased by the Game and Fish Department. Photo by author

Earl Brown

Just below the Hermit Lakes on South Clear Creek were the Troutvale Lakes. Leroy Brown grew up helping his father with the fish business at Brown Lakes:

My father was Earl Brown and he operated the Brown dairy in Ouray. He sold out and moved over here to the Wrights' Ranch in 1898. Then Dad homesteaded the Troutvale Lakes in 1901 and his first wife was Charley Mason's daughter Alice. I was born at Hermit Lakes in 1911. Mabel Wright was my first teacher. The schoolhouse was half way between Brown Lakes and Hermit Lakes.

A man named Burnett built the upper lake and called it Troutvale and Dad homesteaded the upper sixty acres and bought some state school land and then about 1931 he bought the lower lake from the Lariat Ditch Company. Dad increased the height of the upper dam. All the lakes became known as Brown Lakes. We stocked nothing but brook and they reached five to seven pounds. School lake had the largest fish. There were partial winterkills, but I never remember a

Putting up ice each winter was an essential part of the commercial fish business. At Brown Lakes ice was stored in a sawdust-insulated ice house that kept the ice blocks frozen through the summer months. Pictured is Earl Brown, his wife Pearl and unidentified girl (circa 1930). Photo from Janice Nelson Collection

total freeze out. The only thing that kept them alive through the winter was South Clear Creek coming down from that upper country. As the years went by the silt washed down and spread out across the valley above the lakes. When the freeze came the creek was too shallow and the ground would freeze and force the water to spread out all over that upper end and go out on top of the ice. That is when it started to winterkill.

We caught fish by netting. We shipped after June through spawning season into October. When spawning season started then everybody quit shipping to get eggs for hatching to replenish their brood stock. In 1930 Dad and I took 300,000 eggs that we put in our hatcheries and we put another 200,000 in the federal hatchery. We used two hatcheries, one at Hermit and one between Hermit Lakes and Brown Lakes. We lived in the cabin at Brown Lakes for three winters. We sold our fish to the Brent Mercantile in Denver. It was the biggest fish house, but there were several others.

Dad wouldn't let anyone fish in the Brown Lakes except his friends. He kept that strictly commercial, but South Clear Creek was the best fishing creek in the country-all brooks. Suckers didn't come in until the 1940s. I think they were brought in by those guys who seined them

Earl Brown in about 1920 attending to the first fish hatchery located between Hermit and Brown Lakes. The hatchery consisted of a single wooden tank placed into a spring. From Janice Nelson Collection

up and brought them up here to fish with minnows. We never caught any suckers when we were netting. I also worked for Bert Hosselkus over at Road Canyon for three years and I never caught a sucker.

There was plenty of winter work to go along with the commercial fish business. The coldest job was putting up ice. At Brown Lakes, for instance, there were three ice houses built at water's edge. The log ice houses were insulated with sawdust. Before the days of gas engines a single buck saw was used to cut slices of ice that were a foot or more thick. Workers using steel bars chipped fifty-pound blocks that were pulled from the water with large ice tongs and hauled by sled to an ice house. The ice was stacked inside and sawdust was laid between each block on all sides to keep the blocks from freezing together. Ice was absolutely essential for keeping fish fresh for their trip to the market, because there were no refrigerator rail cars.

Bert Hosselkus

Another major commercial fish culturist was Bert Hosselkus who came to the Upper Rio Grande from Quincy, Illinois, with his parents when he was eight or nine years old. He drove a stagecoach, ran a restaurant, delivered mail, and was interested in mining, cattle, and sheep. He got started in the fish business by trading a cow to the Regans for spawning privileges at their lake at the head of House Canyon.

The Lost Lakes north of Rio Grande Reservoir were built for water storage. Bert Hosselkus developed these lakes into a commercial fishery (circa 1930). Photo from Janice Nelson Collection

Hosselkus worked closely with Shrive Collins who homesteaded the Lost Lakes north of Rio Grande Reservoir. Hosselkus started his fish culture business at the Lost Lakes and named his operation after those lakes. His business flourished to such an extent that he built a telephone line up House Canyon to the Lost Lakes so that he could communicate orders for fish to his workers.

Collins built dams on several other natural lakes to develop his water storage rights and sell water to the farmers in the San Luis Valley. Hosselkus expanded his fish business by leasing the fish culture rights on those reservoirs. Collins built the Road Canyon Reservoir in 1908 and sold it to Hosselkus in 1912. Later Hosselkus increased the height of the dam for the Road Canyon Reservoir. He built a house and other buildings to accommodate his fish business just below the present spillway. There was a great cooperative spirit between Hosselkus, Charles Mason, Frank Swank, and local rancher Dan Soward. Hosselkus was an energetic man who sold his fish and eggs all over the

Bert Hosselkus operated Road Canyon Reservoir as a commercial fishery for many years. His family lived in the house across the dam, shown here in about 1956. Photo from Division of Wildlife, George Andrews Photographer

United States and in some foreign countries. He built a hatchery along the Rio Grande just below the Deep Creek bridge in the early 1920s and moved his operation later in the decade.

Bert Hosselkus experimented with Kamloops salmon and steelhead with no success. He is the individual who stocked the first brown trout in the Upper Rio Grande as reported in the *Creede Candle:*

> A new species of trout will soon be introduced by BC Hosselkus . . . receiving 100,000 Lock Laven from the Leadville Hatchery. None of this kind are now in any of our waters . . .

It is unknown where all these brown trout were eventually stocked, but Hosselkus did stock the first brown trout in Lock Laven Lake on the Soward Ranch. Those fish were most likely the initial source of brown trout that populated the Rio Grande in 1927.

Hosselkus successfully introduced the tench, an exotic German species that resembles a carp, but is more fine scaled. He brought those in because they had a reputation for being a good food fish. He stocked them into Road Canyon and into Dan Soward's lower lake. At one time Hosselkus shipped up to two tons of tench every two weeks and as many as seven and a half tons in one month.

The following article in October, 1914 *Field and Farm* magazine described his operation:

Bert Hosselkus was a pioneer fish culturist who built one of the most successful commercial fishing businesses in southwestern Colorado (circa 1940). Photo from Esther Hosselkus Hinshaw collection

> BERT HOSSELKUS FACILITY
> NEAR CREEDE
>
> One of the best fish hatcheries in Colorado is owned by Bert Hosselkus at the Lost Lakes in Mineral county. This establishment spawns 8,000,000 trout fry every year from its hatchery in the Road Canyon near the headwaters of the Rio Grande and pretty well up in altitude. He sells at from $3.50 to $4.00 per thousand and the State of Colorado takes all the surplus of something like two million fingerlings. James Stell of Delta county recently stocked up a new trout run from the Hosselkus Hatchery. His place is at a chain of lakes high up on the Grand Mesa between Cedaredge and Collbran.

By 1914 the Hosselkus Hatchery was the largest trout egg producer of any private trout hatchery in Colorado with an eight million egg capacity. Bert's son Allan "Tuffy" ran the hatchery operation for many years and requested that upon his death the hatchery be sold to Mervin Hartman who operated the hatchery until 1973 when he leased it to the Division of Wildlife. After the Division's lease expired in 1978, Charles and Jane Downing bought the hatchery and operated it as a

The Wetherill fish hatchery was small, but produced millions of brook trout. The hatchery was in the lower level of the building. Photo by author

commercial hatchery until the late 1980s. Since then the hatchery has not been used for commercial fish and egg production, but is part of the Miners RV Park. The Hosselkus operation was the longest operating commercial fish business in the Upper Rio Grande.

The Wetherills

Clayton Wetherill, brother-in-law to Charles Mason, came to the Upper Rio Grande soon after Mason. He bought the land between the San Juan Ranch and Highway 149 from the Officers family. The stone foundation and original building are still in place between the San Juan Ranch and Highway 149. Clayton Wetherill died in 1921, and his wife Eugenia bought and developed what is now the Wetherill Ranch.

Hilda Kipp, Carroll Wetherill's sister, described the hatchery:

> It was just a small hatchery. The hatching troughs were built at ground level below the framed portion of the house. There were 12 troughs and a spring was piped down to the house and water ran year around. There was a tank that was hand dug and boarded up inside

and then there was another one behind the stone house and it had baffles in it and it was twelve feet long and 8 feet wide and the water was an ideal fifty two degrees year around. Spring Creek Pond, just below Ptarmigan Meadows was initially built to supplement the spring water for the hatchery.

Howard Kennell who was born and raised in Creede and Saguache said:

When Carroll and I were about seventeen years old we ran the hatchery for two years. Every fall we took a million brook eggs at Wright's lower lake next to the headquarters. Brook trout was the only fish we hatched in that hatchery. We only hatched eggs and fry. Carroll and I built the middle dam on Wrights' Ranch with "V" handled shovels and a team of white horses. Wallace Wright built a check dam there so we could trap brook that were running about two pounds and then they got to winterkilling. Wallace got the idea that if he built an upper dam and released water from it to the lower lake that it wouldn't winterkill, but that never worked.

Carroll Wetherill, the son of pioneer Clayton Wetherill, was a rugged mountain man. As one of the San Luis Valley's premier trout growers he was dedicated to improving the fishing in the Upper Rio Grande. Photo from Carol Ann Wetherill Getz Collection

The Wills on Crooked Creek

Frank and Shirley Wills were brothers who homesteaded on Crooked Creek. Shirley built the S Lazy U Ranch, which had a small hatchery. Frank built the Cliff Ranch. After Frank and his wife divorced, their son Billy ran the Cliff Ranch Resort. These ranches became resorts that catered to fishermen who enjoyed fishing in their lakes. The fish raised in this drainage were harvested by fishermen rather than being marketed to restaurants and grocery stores. Wills built a fish hatchery near the confluence of Miners Creek and the Rio Grande in the early 1960s. Over the years he sold millions of fry to area resort owners.

Lawrence Phipps

Lawrence Phipps bought several homesteads that collectively became the La Garita Ranch in 1908. He was an industrialist who became a United States Senator from Colorado. Whether some of the nationally prominent conservationists and pioneer conservation organizations of his day influenced him to become a leader for conservation in Colorado is unknown, but he sponsored a number of conservation projects including the Creede federal fish hatchery. More than this, he practiced conservation on his La Garita Ranch and began a "legacy of stewardship" that his sons and grandsons continued.

The La Garita Ranch had its own fish hatchery. Its operation was different than most of the other private hatcheries in the region. The hatchery building was twenty by thirty-two feet and had nine troughs that could handle about 40,000 eggs each. There were no screens on the lakes, so fish could come and go as their genes directed them. After spawning season, the mature fish were released to a ditch that went down to Bellows Creek and many found their way to the Rio Grande and were enjoyed by many fishermen. This was done by design. The senator was a man ahead of his time. The *Creede Candle* said of Lawrence Phipps:

> Senator Phipps spends a portion of each summer season at the ranch when he usually fishes Bellows creek or the river part of each day. He uses barbless hooks always when fishing and as he often has guests who come to enjoy the privileges of fishing on his property while he does not attempt to dictate, he usually lets them know he would like to have them follow his example and use the same kind of hook, which do not inflict as much damage as barbed hooks to the small fish that take the fly or bait and must be returned to the water.

Colonel A.E. Humphreys

Colonel A. E. Humphreys visited the Hot Springs Lodge at Wagon Wheel Gap in 1896. Humphreys was a wealthy man. He liked the country enough to buy land where he built a lodge and other buildings. In the 1920s he built the only concrete dam in the Upper Rio Grande which created Humphreys Lake on Goose Creek. He also built Haypress Lake. Although Humphreys built a small fish hatchery for stocking his lakes, he never participated in the commercial aspects of fish production. However, he was a leader who wanted the fishing to be as grand as it had ever been. In an interview reported in August 13, 1922 *Rocky Mountain News* Humphreys stated:

Colonel A.E. Humphreys was the first vice-president of the Izaak Walton League. He was a pioneer conservationist whose passion was to improve fishing in Colorado. Photo from Ruth and Darcy Brown Collection

> I will stock only the cutthroat-no brook. I shall raise enough to stock Goose Creek and the Rio Grande. I will put in a million fry at once and am planting fresh water shrimp, which thrive near Del Norte, in the lakes to feed the young trout. This shrimp multiplies, they say and fish love it. I will have my own hatchery. It is built and ready. My nursery lake will be the next step-then the two big lakes-Hay Press and Humphreys, where millions of trout will live; thousands are sure to go over the spillways into Goose Creek and the Rio Grande, which is right for I want to see the trout streams of Colorado again in the glory they had before they were fished out. I intend to help.

Humphreys was a strong supporter of the development of stream-side ponds along the major rivers. He donated $250,000 to the Game and Fish Department for this purpose. His greatest influence on the improvement of fisheries nationally and in Colorado, however, was through the Izaak Walton League.

Lost Lakes and Seepage Lakes

The Lost Lakes, north of Rio Grande Reservoir, have been a private fishery since Shrive Collins homesteaded and built dams on the lakes in 1908. Howard Kennell recalled:

> When Shrive Collins first developed the Lost Lakes he leased them to Bert Hosselkus. Later that lease was taken over by Ted Walker, the caretaker at the Farmer's Union now called Rio Grande Reservoir. Walker changed the operation. Two of the conditions of his lease were

to take fishermen in there and keep the lakes stocked. I made a deal with Walker and I took eggs mostly rainbow and natives from 1929 to 1934. Walker also built the resort at 30 Mile Bridge that is now called Trego's 30 Mile Resort below Rio Grande reservoir. He had a string of horses. I did a lot of packing for him and took dudes over to the Pine, Flint and Ute Lakes. Wallace Wright and Billy Wills also had the lease for a few years. After World War II Carroll Wetherill took over the lease and ran it for many years.

Shrive Collins also built dams on Seepage Lake in 1908 to provide water for his ranch north of Del Norte. Bert Hosselkus leased these lakes for commercial purposes for a short time, but Melvin Powell, who built what is now the Broken Arrow Ranch, used the lakes for his guests. George Wintz said:

Seepage lakes had the best water for raising fish. It was warm and even in the winter never froze over completely. In a year or two small fish would grow to be large. The lakes are controlled by the depth of the water in the Santa Maria Reservoir. When it gets up so high it seeps underground and surfaces to form the Seepage Lakes. In those days (1908 to the 1930s) they hardly ever went dry, but in drought years they dry up completely. I stayed up there one fall and trapped muskrats. I sold them to Al Birdsey for a dollar a pelt and he sold them for two dollars. A dollar in them days could buy about ten dollars' worth now. We would kill a duck now and then. We used to stay in an old cabin there for up to ten days and trap muskrats. One year we caught 120 rats. We did it to save the dams from being tunneled through by the rats and then having the dams wash out. I just had a lot of fun and made a few dollars too.

Soward Ranch

There were ranchers who saw the fish business as another way to increase their cash flow. Dan Soward built a fish hatchery and several lakes on his ranch at the Soward Ranch about 1910. Emma McCrone recollected:

The Soward Ranch allowed no fishing. That was a terrible policing job of keeping people out of our lakes. My father had a crusty dispotion that kept them away. We had a routine to market our fish: We set the gill nets on Saturday and Sunday. Monday we pulled the gill nets. I have dressed many a trout. We'd put them in the wash tubs and we had to get them dressed, boxed, and iced by Monday and down to Creede to go out on the train at 4 o'clock in the afternoon. Every winter we put up about 20 tons of ice for the Soward Ranch fish operation.

We shipped to the Antlers Hotel in Colorado Springs, the Brown Palace Hotel, the Broadwell restaurant in Denver, and another restaurant in Trinidad. When you sold to a restaurant you sold three fish to the pound, because that is what the cook is going to serve you. We sold the bigger fish to a big grocery store in Denver called the Hurlbert. They would take any fresh fish. Housewives weren't as particular as the restaurants.

Father tried steelheads that Bert Hosselkus gave him, but they didn't work out. Then one day Bert, who was a good friend, brought in some brown trout and asked father to experiment with them. So they were stocked into Lock Laven Lake on the banks of Trout Creek.

The brown trout did well in the lake until a flash flood in 1927 changed the Rio Grande fishery forever. The *Creede Candle* reported on July 2, 1927:

RIO GRANDE OUT OF CONTROL, LOSS IS REPORTED HEAVY

Due to the heavy rains in the mountains the waters of the Rio Grande and all tributaries in this section have been swollen out of their banks. By Tuesday evening [June 28, 1927] bridges were under guard and no trains arrived after that day.

Three of Dan Soward's well stocked fish lakes have gone out with a big financial loss to the owner. Deep Creek bridge is probably a total loss. The approach to Five Mile bridge will have to be replaced . . . Squaw Creek bridge, Art Neale's bridge, the Gap bridge, Red Mountain and Trout Creek bridges were all lost or badly damaged.

Roy Powell, whose family homesteaded the Powell Ranch on Trout Creek upstream from the Soward Ranch, remembered that the brown trout came from Soward's Loch Laven Lake on Trout Creek. During the 1927 flood, the creek washed that lake out and that is

The Soward Ranch where Loch Laven Lake is now a small, willow-filled depression next to Trout Creek in Antelope Park. It was the site where the brown trout was first stocked in the Upper Rio Grande. Photo by author

when the browns got started. Brown trout were not seen prior to that flood. Before then Trout Creek had mostly native cutthroats and some brook trout and the river also had some rainbows.

The very next year, in the fall of 1928, the San Juans were experiencing the other climactic extreme of a severe drought. Hilda Kipp said that her mother called from the Wetherill Ranch to Pat Dumont at the Del Norte Hatchery and told him that the Rio Grande had dried up to such an extent that there were places where fish were trapped in pools that had no water flowing. Fish were going to die if they weren't seined and moved to flowing water.

The *Creede Candle* reported:

Land-locked Fish Returned to River

Thru the efforts of A.L. Dumont, superintendent of the Del Norte fish hatchery, J. Whitmore, his assistant, and a number of volunteers from Creede, saved thousands of trout measuring from six to fourteen inches in length. They were salvaged this week from sloughs in the vicinity of Wrights' Ranch and returned to the river near there.

> The sloughs were formed by a change in the course of the river when the stream got low, and if they had not been seined out and turned into the river the fish would have been frozen when severe winter weather comes.
>
> It is estimated that the number of fish salvaged by the work was around 4,100 and as the probabilities are they will survive in the stream, making better fishing next season. This is the first time such a transfer has been made, it is said, and credit for the idea is given the officers of the local chapter of the Izaak Walton League, which was enthusiastically supported by Supt. Dumont who took charge of the work. The local men who assisted were Victor Swanson, Brose Mortensen, Lou Bruns and Wm Jackson. The same plan to salvage a good many more trout near Wagon Wheel Gap also was carried out by Superintendent Dumont, at a place where the fish had been trapped away from running water by a change in the course of the river.

Both flood and drought have impacted the fishing in the Upper Rio Grande over the years. In spite of all the setbacks that nature has thrown at human efforts to manage fisheries, the effort to improve fishing has never slowed. The private hatcheries began as commercial operations to market fish for food. The role of the private fish hatchery began to change after World War I when the automobile brought more anglers to fish in Colorado. They began to supply fish to the private lakes and streams of ranches and resorts, because the state hatcheries would not stock waters that were closed to the public. Some of the hatcheries were very small and were only intended to provide fish for a local stream or lake, while others became major businesses. All in all they met a need, but over time most of the old operations have gone out of business, though some still provide a catch pond for local resorts.

Commercial aquaculturists are making a comeback in southwest Colorado, including the San Luis Valley. There are several very successful fish raising businesses that market their food fish. Some high schools are offering aquaculture classes to teach students how to raise fish for commercial purposes. There is a market for commercial fish once again and entrepreneurs will rise to the occasion to meet the need.

Chapter 4

THE STATE AND FEDERAL HATCHERIES

When automobiles became more common, people began traveling farther from home to enjoy the great outdoors and were soon catching more fish than natural fisheries could produce. This increased tourism brought more money into Colorado and businessmen began to promote the state's magnificent scenic, fishing, and hunting opportunities. Some of this increased demand for fishing was caused by residents of Colorado sending over 500,000 picture postcards in May of 1911 advertising the marvelous fishing in the state. The Denver Chamber of Commerce provided the postcards to residents to mail to their friends and relatives to entice them to come to Colorado for fishing vacations. Local resorts often provided their guests with local postcards with beautiful scenery and successful fishermen to further entice their friends. The seeds were successfully planted to increase the number of fishermen coming to Colorado.

There was never any question that the only way to provide more fish was to raise them artificially. For hundreds of years, in Europe and Asia, artificially raised fish had supplied the demand for fish both for sport and food. Federal, state, and private hatcheries were already being built all across America. Following the tradition of this success, it was only natural for sportsmen to demand that more fish hatcheries be built in Colorado.

The Earliest State Fish Hatcheries

Distribution of fish was difficult because roads were poor and fish were transported by horse-drawn wagon. Trucks and cars were just coming into being. Therefore it was necessary to have hatcheries scattered around the state so they would be close to the waters that needed to be stocked. The first fish hatchery built by the Game and Fish Department in Colorado was built in 1893 near Hermosa, north of Durango, and was operated for ten years. The Durango hatchery was built in 1903 and has operated longer than any other state hatchery. Highways, vehicles and the equipment necessary to transport fish were very limited.

The first fish hatchery in the San Luis Valley that was built by the State of Colorado was on San Francisco Creek, south of Del Norte, in 1908. Soon after construction there was an explosion and fire that destroyed the hatchery, but it was rebuilt. This hatchery stocked fish primarily in the western portion of the San Luis Valley. The Del Norte hatchery had very limited capacity, because of an inadequate water supply. In 1915 Walter Fraser, the Colorado Game and Fish Commissioner, wrote in his biennial report to the Governor:

> The Del Norte hatchery site should never have been acquired as results were less than satisfactory, but a new pipeline was put in to enhance a spring. The Department recognizes the Rio Grande Fishermen Association who so liberally contributed funds and labor and now this unit should rank favorably with other units.

A state fish hatchery was built in Antonito in 1914 and operated through 1929. The La Jara hatchery was built in 1933 and was closed in 1977. These hatcheries furnished fish to the Conejos River drainage. The lack of suitable water, and in the case of the La Jara hatchery, the expense of pumping water, made raising fish too expensive.

The Creede Federal Fish Hatchery

Since the creation of the U.S. Fish Commission in 1871, Congress had on various occasions entertained legislation to finance federal fish hatcheries across the country. The Leadville, Colorado federal hatchery, built in 1899, was the earliest hatchery constructed west of the Mississippi River. The desire for a federal fish hatchery for the Upper Rio Grande was expressed at the first meeting of the Rio Grande Sportsmen's Association in 1910. It was apparent at that time that fish would have to be raised locally, as transportation was so limiting. No action had been taken on most fish hatcheries when the priorities of World War I put much federal spending on hold until war's end in 1918.

The earliest citizen group to focus its attention on the nation's fisheries was the Izaak Walton League, formed in 1922 at Chicago, Illinois, with William Diulg as president and Col. A.E. Humphreys of Wagon Wheel Gap as the first national vice-president. The Izaak Walton League was nationally recognized with more than 250,000 members in 3,000 chapters across America. The first Colorado Chapter was started in 1923 at Denver with Col. Humphreys as the first president. Creede started its own chapter in 1925.

Since federal legislation was necessary for funding the Creede hatchery, it was important to gain as much support as possible throughout Colorado. Bert Hosselkus, one of the local fish culturists in the area; A.H. Wason; and others from Creede attempted to influence other sportsmen to support the Creede hatchery. For example, in August of 1925, they visited a number of towns on the Western Slope and attended a meeting of the Federated Sportsmen's clubs at the Grand Mesa at which meeting they presented the plan for establishing a Federal Fish Hatchery on the Rio Grande. They secured the endorsement of that group.

Hosselkus and A.H. Wason prepared a formal application for a federal hatchery on the Upper Rio Grande to present to Congress. They had a meeting at Col. Humphreys' resort near Wagon Wheel Gap and while there secured the endorsement of the Colonel and of U.S. Senator L.C. Phipps.

Bert Hosselkus traveled to Chicago to lobby for the Izaak Walton League endorsement as reported in the April 11, 1925, *Creede Candle:*

> B.C. Hosselkus reached home yesterday after ten days absence in attendance at the national convention of the Izaak Walton League in Chicago, which he attended as a delegate from the Creede Chapter.
>
> The important effort in addition to the duties of a delegate, was that of securing the passage by the convention of our resolution for a federal fish hatchery in this territory. Nine resolutions for such hatcheries were before the convention, seven of which failed in the committee. Two of them reached the convention floor, and then there was but one, our own, and it was endorsed without a dissenting vote.
>
> Much credit is due Mr. Hosselkus for his work in this matter. And to those who unselfishly supported the movement, its successful termination is truly gratifying. Fifteen hundred sportsmen were in attendance from every section of the United States.

By the winter of 1926 the slow turning wheels of government were beginning to respond to its citizenry. There was popular support throughout Colorado and especially from the southern part of the state. Senator Lawrence Phipps worked with Congressman Guy L. Hardy, who presented a bill to Congress to authorize an appropriation of $50,000 for the Commissioner of Fisheries in the Department of Commerce to build a federal fish culture station as an auxiliary to the Leadville federal hatchery. Soon the Federal Fish Commissioner and a delegation from Washington, D.C., met with Bert Hosselkus and others in Creede to select a site that had the necessary water conditions for a trout hatchery.

The site under consideration was about two miles below Creede along the banks of the Rio Grande. The site was easily accessible by automobile all year. The new hatchery would be the only federal hatchery having the convenience of a railroad spur, so it would not be necessary to haul shipments of eggs and fry between the hatchery and fish railcars. It was the intent of the government to make the new establishment one of the principal ones from which to supply federal hatcheries in other parts of the country. The proposed plan was to drill a well that would provide water at a constant forty-two degrees Fahrenheit and use an available spring which was fifty-two degrees, thereby making it possible to regulate the water temperature at will, retarding or advancing the development of eggs and fry to suit the demands of the season. The optimum temperature for hatching trout eggs is about fifty-four degrees Fahrenheit.

Community excitement grew when it was announced in the August 6, 1927 *Creede Candle* that the site was selected and the legal processes started:

> **Federal Hatchery Site is Approved**
>
> Assistant Commissioner Leach arrived from Washington, D.C., Thursday to go over the ground that had been recommended by Mr. Van Atta, of the Bureau of Fisheries Leadville, as a site for the new Federal Hatchery. The spot chosen for this covers 30 acres just east of Willow Creek on B.C. Hosselkus' property, which will be presented to the government by Mineral county.
>
> After a careful inspection of this ground and conditions surrounding it, a trip was made to various spawning beds, including Continental, Santa Maria and Humphreys Reservoirs. Since all other Colorado sites have been eliminated it seems most probable that Mr. Leach's

> unreserved approval will precede the early signing by the Federal Fish Commissioner of papers which will bring the hatchery here.

There were details to work out such as exchanging road access across Dr. Fahnestock's Wason Ranch and road and pipeline easements across Bert Hosselkus' property in trade for electric light and power supplied by the City of Creede hydroelectric station.

By the fall of 1929 all the plans, materials, and manpower were lined up. Timber had been cut near Leadville and milled at a sawmill at the Leadville hatchery. The lumber was hauled to Creede in Army trucks. William Whitlock of Colorado Springs was the contractor. Superintendent C.H. Van Atta of Leadville and his assistant H.J. Whiteman supervised construction.

The stock market crash of 1929 had just occurred and the Great Depression gripped the nation. Jobs were almost nonexistent in the Upper Rio Grande. This project, small by today's standards, provided work for many Creede men. In mid-November the first manager of the Creede Hatchery was announced in the *Creede Candle:*

> Charles Fuqua has taken charge at the new federal fish hatchery that is being constructed on the Rio Grande below Creede, and he will be the foreman of the hatchery after it is completed and put into operation. Mr. Fuqua comes from a government hatchery at McAllister, Montana where he was fish culturalist in charge of the station.
>
> It is expected the main hatchery building will be finished about December 1st. Work is underway on the garage and construction of the cottage where the foreman will have his residence also has been started. The main building is being built under contract, but the other construction work is being done by employees of the Bureau of Fisheries and a number of local men.

In mid-November of 1929 the Creede hatchery was accepted. The new hatchery had one main building with outside dimensions of thirty-six feet by sixty-five feet, with most of the first story arranged to accommodate the hatching troughs of which there were forty-eight, in two rows of twenty-four each. The hatching troughs were shipped from the Leadville hatchery. This number of troughs had the capacity for handling ten million eggs, or two to three million fingerlings.

The Creede federal hatchery in about 1954: hatchery building, ice house, workshop (above), and residences (below), with ponds that were operated from 1930 to 1965 by the U.S. Fish and Wildlife Service. The hatchery facility is being modified as a museum and a hatchery for Rio Grande cutthroat trout. Photo from Dot Wilkerson Collection

The second story of the main building was arranged for living quarters for several employees, or one employee and his family, and was divided into three bedrooms, a kitchen and a bathroom. At the east end of the building a hot water heating plant was installed, and a room provided for use as an office by the foreman.

The garage accommodated four cars, and at one end was the grinding room, where

liver and other feed was prepared for feeding the young fish. In the second story were carpenter and paint shops. The foreman's residence sat up on the mesa east of the hatchery. The building covered an area twenty-nine by thirty-eight feet, and had six rooms and a bath, with all other modern conveniences. Later on when more money would become available, another cottage for the use of employees was built near the foreman's residence.

The main building was built in two months for $12,000 at below the estimated cost. The amount of $30,000 was appropriated for the entire enterprise, including hatching troughs, pipelines, and other equipment.

The first fish eggs hatched in the Creede hatchery were furnished by the Mason hatchery at Hermit Lakes. Mason supplied more than a million brook eggs in the fall of 1929. By the spring of 1930 the new Creede hatchery was preparing to stock its first fish. In mid-May the first trout were planted in Squaw Creek and the Rio Grande Reservoir. The hatchery did not have a truck to haul the fish cans so Mineral County provided one. Volunteers from the Creede Chapter of the Izaak Walton League, including Frank Powell, Charles Spitz, and E.E. Banakin, planted the tiny trout, which were each about an inch long and were carried in twenty-three cans.

The Continental Reservoir was set aside as a spawning refuge for taking cutthroat eggs. In mid-May of 1930 hatchery personnel Floyd Roberts and Harry Larson were taking native cutthroat spawn at the Continental Reservoir. At that time of year there was still deep snow and the ice had not melted. They had to pack in all the egg-taking equipment such as seines, nets, egg boxes and dish pans on horseback. In a couple of weeks when the ice cleared they used a motor boat to carry supplies back and forth as needed. They expected to take a couple million eggs.

The hatchery didn't have a budget to buy eggs so hatchery personnel found places to take spawn from wild fish. Taking eggs from wild fish was often a very miserable job, especially at a deep lake such as Lake San Cristobal which normally freezes over in late fall or early winter before the fish are ripe for spawn taking. Ivan Vanaken worked at the hatchery and vividly remembered taking brook trout spawn:

> Brook trout congregated around springs at Lake San Cristobal and could be seined with great effort. We chopped several holes in the ice

surrounding the spawning bed. We pushed the seine under the ice using a long pole and then caught the seine and repeated the process from hole to hole until the seine circled the fish. Then the seine was pulled in and the fish caught and put into live boxes until after a few days' efforts there were enough fish to spawn to make it worthwhile. We had to chop the holes open every day, because they would refreeze. I remember spawning brook trout when it was twenty-six degrees below zero. We would heat a kettle of water and occasionally dip our hands and arms to keep them warm. The last year we took spawn we put in our report how cold it was and we got a letter from Washington canceling that spawning operation, because it was too cold to have men working under those conditions.

From two to three million brook eggs were taken each season from some twelve thousand females. The eggs were transported to Creede in ten gallon oak kegs. About ninety-seven percent of the eggs eyed-up which was a very high hatching return.

The Creede and Leadville hatcheries started a rainbow trout egg-taking operation at Eagle Nest Lake in northern New Mexico. Hatchery personnel built an egg-eyeing station to incubate the newly taken green eggs to the eyed stage. Eggs transported at this stage of development had a much higher survival rate during transport than green eggs. Eggs were then shipped to Creede and other federal hatcheries.

Commercial operators also furnished eggs to the Creede hatchery and payment was made by the hatchery keeping a percentage of the hatching eggs and the private vendor selling his portion. There was a lot of hard work invested to provide those fish eggs and sometimes it became a real adventure. According to Howard Kennell:

Charles Fuqua was the first superintendent of the Creede hatchery and in 1931 he didn't have money to take eggs so he initiated a deal with Carroll Wetherill and me to let us take eggs on a share basis at Emerald Lake over on the Pine River. From the old records he thought we could take several million eggs. You see, the state had a program at Emerald Lake years before. They had been taking quite a few eggs from the Bayfield side. Charlie initiated the deal and we went along with him, because we could sell our share of the eggs to Bert Hosselkus. He could handle all the eggs we could bring into his hatchery.

The Creede federal hatchery personnel built a fish egg spawning station at Eagle Nest Lake in New Mexico. John Thompson (below), superintendent, taking spawn in about 1934. Photo by DC Booth (National Historical Archives).

Charlie loaned us a good amount of federal equipment for taking eggs. We were just kids seventeen and eighteen years' old. We had our own saddle horses and we borrowed mules that belonged to Shirley Wills. We rode horseback about twenty-six miles from the Farmer's Union dam over Weminuche Pass and down the Pine River all the way to Emerald Lake.

Carroll and I took four egg cases and all our gear and we set the trap and nets. When we got one case about full of eggs, we put them in a snow bank. We stored them in the old wooden egg cases packed with peat moss and cheese cloth with an ice tray on top that would drip cold water onto the eggs. They were the size that would fit in a pannier so you could pack a case on each side of a horse or mule. It was quite a delicate process to take care of the eggs and get them back to the hatchery alive.

We were staying in the cabin between the two Emerald Lakes. We seined the fish in the lake and set a fish trap in the creek. Well, we had the trap full of fish, but the fish were not ready to spawn and we were going to wait a day or two and we got up one morning and ate breakfast. The cabin had a dirt floor and a bed with springs in it and we had our own bedding and we had kept some male fish to eat. We always packed plenty of food on a trip like that, but we also ate fish.

There were still two fish left on a plate when two guys rode up as we were just getting

ready to take eggs. One guy was a game warden out of Bayfield named Moss and the other was a ranger was named Norris out of Durango. They had heard through the grapevine that we were up there and they were going to put the skids to us, because they didn't want us over there in what they figured was their country.

The ranger got off his horse and spoke to us and came into the cabin and asked, 'What are you guys doing over here?'

I said, 'We are taking eggs.'

He asked, 'What are you doing with the eggs?'

I said, 'We are taking them over to the Federal Hatchery in Creede.' He acted like there wasn't a federal hatchery at Creede.

He said, 'You know I think you guys are lying and you are stealing these eggs.'

I said, 'No, we're not stealing eggs, we are taking them to the hatchery.'

"He asked for credentials, and a permit to take eggs. We never had anything in writing. We're just kids and we don't have anything in writing.

'Yeah, you're stealing eggs!' he said. 'What do you have up here anyway?'

I said, 'Well, we got a big seine trap set on the big lake.'

'Well', he said, 'go get your horses and we'll go take a look.' He made us pull that seine in and turn all the fish loose.

He asked, 'Where's the rest of your fish?'

I said, 'Up in the creek in a trap.'

So before he got out of the cabin he saw the fish in the pan and he asked, 'You kids got a license?'

We looked at one another and asked, 'What kind of a license?'

'A fishing license.'

'No, we don't have a fishing license.'

'I'm going to take you to town . . . turn them fish loose, and straighten this thing out.'

So after we pulled the seine to let it dry out and then pick it up on the way back. Carroll and I looked at each other and thought, . . . we're going to jail! . . . and so we got down to the cabin and got all our stuff loaded on the mules and in the process I told him that the mules didn't belong to us and he asked, 'Who did they belong to?' I told him they belonged to Shirley Wills who ran a dude ranch over by Creede on Crooked Creek.

'Well' he said, 'We'll just hold this livestock, until your fine is paid.'

So those guys rode on ahead of us and every so often they would stop and talk. Carroll and I were in the back and when we'd stop we'd talk about what they were going to feed us in jail and what are we going to do about the mules and what are we going to tell Shirley about the mules? We were in a hell of a fix.

When we got down to the Pine River the ranger got off his horse and I rode up to him and he said, 'Well boys I'll tell you what we are going to do with you. We have changed our minds and going to let you take the mules back if you promise us that when this comes to trial you will come over.'

I said, 'You bet we'll gladly come over.' We were glad to get out of that. So they turned us loose with all that gear and they let us keep a full case of eggs. So we rode all the way up the Pine River and down the Weminuche to the reservoir that night about ten o'clock and we walked down to Walkers 30 Mile Resort and called on the telephone to Charlie Fuqua and I told him that we had got run out.

'What do you mean you got run out?' he asked. I told him we got some eggs if he wanted to come up and get them and he said he would send a man up and they came up in a little Model T truck and picked them up. We left the mules and horses at the dam and we drove our Model T down to the hatchery and he was madder than a hornet. He said, 'You kids did a good job.' And the more we told him the madder he got.

I told him, 'Charlie, I showed him the egg cases with the U.S. Bureau of Sport Fishery insignia. He wouldn't acknowledge it at all. He accused us of stealing it all.'

Charlie said, 'I'll get those two guys.' He said, 'You guys get ready to go back over there.'

We decided we didn't want to get into that mess again. We went back to the Wetherill hatchery and told Charlie we wouldn't go back and by then it was the middle of June and the fish would be finished spawning anyway. We never heard another word from the ranger or warden. Oh what we went through in those days to raise fish.

In 1930 work to improve the hatchery began with the advent of the new fiscal year. Only a few men were put on at the start, but in a month or so the work force was increased. Local men completed the residence of Foreman Fuqua and built the pipeline across the Hosselkus property for an increased water supply from the big spring above the Deep Creek Bridge.

Most of the fish stocked in the early years were small fry. These tiny fish had a very high mortality rate when stocked into the wild, especially if the recipient waters already contained larger predatory trout. The Colorado Game and Fish Department was just beginning to stock catchable-sized (eight-inch fish) using small trucks equipped with steel tanks. Charles Fuqua announced a major change at a 1930 meeting of the Creede Chapter of the Izaak Walton League as reported in the February 1, 1930, *Creede Candle:*

> Charles Fuqua and John Harrington, his assistant, were present and gave interesting information concerning the matters discussed. The Rocky Mountain states are the only ones that continue to plant fry in streams. It now is the policy of the Bureau of Fisheries to plant only larger fish for stocking fishing waters.

The daily routine at the hatchery varied with the seasons. In the spring there were rainbow and cutthroat trout spawning and those egg operations required removal of the dead eggs to prevent fungus. A number of Creede women got temporary jobs picking eggs. Also, fingerling brook trout that had been hatched from eggs taken the previous fall now required feeding along with daily cleaning of troughs and raceways. There were fish to be stocked in area streams and lakes as soon as the ice was out, and in larger streams when the spring runoff had subsided. At the hatchery, men labored long hours to prepare fish food. Using a cement mixer they blended ground beef spleen, liver, cottonseed meal, wheat midlings, fish meal and other ingredients. Carp were also ground and cooked with oatmeal for feed at some hatcheries and retaining ponds as early as 1932. Such feeding became the norm at state hatcheries in 1938 after the advent of processed dry carp meal made at the state's carp refinery plant at Las Animas. During the long winters the coal-fired furnace had to be stoked to keep the hatchery building from freezing. Many Creede men worked at the Creede hatchery in its thirty-five year history.

State fish hatcheries in Del Norte, La Jara, and Monte Vista stocked many fish in the Upper Rio Grande, but the Creede hatchery produced most of the fingerling trout that were stocked in the back country of the Upper Rio Grande. John Harrington and Gene Mason were superintendents of the Creede hatchery after Charles Fuqua was transferred. Gene Mason was an adopted son of Charles and Anna Mason of Hermit Lakes. Mason learned the fish business from his father. After serving as superintendent of the Creede hatchery he became the superintendent of the Leadville hatchery. Mason's career led him to become one of the distinguished fish culturists in the U.S. Fish and Wildlife Service. At the height of his career he coordinated an international team of fisheries experts for lake trout recovery in the Great Lakes.

According to Carroll Wetherill:

Gene Mason in 1935, the son of pioneer fish culturist Charles Mason, standing at the door of the Creede hatchery. Mason became the superintendent and began an outstanding career as a fish culturist in the U.S. Fish and Wildlife Service. Photo from James Mason Collection

> The role of the Creede fish hatchery became very important in the 1930s and 1940s when the fishing pressure got so much that they had to start stocking the back country streams and lakes. Forest Rangers like Vanaken and Franklin applied for fish from the Federal Hatchery after 1930. After Franklin quit, the local people helped pack fish into the back country. Tom French was then the superintendent and he told everybody that if we didn't get some applications in they would close the hatchery down. The next year everybody applied for fish. There would be four or five applications for one stream and none on another so they set up a system to handle applications and that is how the Upper Rio Grande Fish and Game Association got started. Later we got together with Bill Schultz, Tommy French, and Carl Welch, who had become the first state fish manager in the San Luis Valley,

and put together a coordinated request for fish for the entire Upper Rio Grande. They thought that was okay and our application was sent to Albuquerque and then to the Game and Fish Department in Denver and back to the Creede hatchery.

Stocking fish was not always as simple as it may seem. Sometimes it can be very frightening such as the time I had stocking fish up at Vallecito Lake at the head of Vallecito Creek. I decided that we ought to put some fish in there. Well the hatchery had some fish slated for Ute Creek, but instead I took Doc Barksdale with me over the Divide from Bear Town and down to Vallecito Lake. There was no trail. About two thirds of the way down we had to lead the pack horses across a steep-sloping quartzite outcropping and I didn't think anything about it. We put the fish in the lake and started back up. It started raining and we worked and worked and we couldn't get those horses over that short space of slick rock. We tried pulling them up with our saddle horses and that wouldn't work so finally we cut up saddle blankets and tied to the horses hooves and those horses seemed to know what we wanted and tested the traction and they went right up. I love that country. It's about my favorite place.

The Creede hatchery enabled not only the government but the citizens to improve recreational fishing. From the time of the initial desire for an Upper Rio Grande hatchery in 1910 to its full-fledged operation in 1930, the Creede hatchery was a part of the local community. It not only provided much needed employment during the Great Depression and subsequent years, but it also helped meet the demand for better fishing for a growing tourist industry. The hatchery superintendents were professional fish culturists who applied state-of-the-art equipment and techniques of their day. Its personnel often became leaders in fish culture through the U.S. Fish and Wildlife Service. Even after the unit was officially closed by the Fish and Wildlife Service in 1965 the Division of Wildlife continued to use the facility for hatching cutthroat eggs and for research. In 1999 the Division of Wildlife provided a "Fishing is Fun" grant to create a partnership with the City of Creede to develop the facility into a working hatchery and museum. The unit will be used seasonally to raise Rio Grande cutthroat and Colorado River rainbow fingerlings in an environment free of the Whirling Disease parasite.

CHAPTER 5

RECOVER SOME AND LOSE SOME

By the 1930s elk and deer herds were beginning to increase throughout most of western Colorado. Those species that had economic and sport value began to respond to the protection and attention that man gave them. In some areas the herds had increased to such a level that ranchers were complaining that there were too many animals competing for food they felt was just for cattle or sheep. Sportsmen and a growing urban population wanted even more animals on the open ranges.

During this same time other species that had been hanging by an ecological thread would soon be extirpated from southwestern Colorado. Woolgrowers and cattlemen pressured the government to exterminate all predators in an attempt to reduce their predation losses. Government and private trappers began their task and soon several species of furbears such as mink, black-footed ferret began to disappear from the Upper Rio Grande. Nevertheless, species such as the gray wolf, grizzly, bald eagle, peregrine falcon, were about to be extirpated.

People lived off the land but wildlife was not exploited as it had been in the 1800s because there were laws to prevent exploitation and another character was on the scene to protect wildlife -the game warden. Yet people used hard times as incentive to get ahead. Although the resort business had already begun it increased during the 1920s and '30s as a means to increase ranchers' cash flow. They built fishing camps along the major rivers in Colorado to cater to the increasing number of fishermen.

ELK

By the 1930s people were seeing elk regularly. Three decades of protection along with the introduction of elk from Wyoming had increased the population throughout western Colorado. But elk were not scattered across the San Juans. Rio Grande elk and San Juan elk herds were expanding back and forth across the Continental Divide, but elk had not migrated northward. For example, George Wintz observed that there was the elk country south of the Rio Grande and the deer country on the north side. It wasn't until the late 1930s that elk were seen throughout the San Juans.

There were several factors affecting elk habitat and their migrations. These included the introduction of non-native elk from Wyoming that were released to the south side of the Rio Grande; changes in grazing, logging, mining, hunting, and all the associated disturbances caused by these activities. Ranchers often saw elk, because they were out and about tending to their cattle. Emma McCrone said:

> The first elk I remember were in our cattle on the Soward Ranch. I remember the cowboys coming in and talking about seeing elk grazing with the cattle. They weren't very wild in those days. There was one we called "Old Box Car" for he had a tag in his ear and was one of the original Wyoming elk brought here in 1911. Somebody killed him after the season opened in 1938.

Sometimes cowboys were able to see elk up close and personal such as Charles Dabney's encounter with a big bull:

> I was herding cattle up Rito Hondo Creek in about 1936. Elk hadn't been hunted yet and they weren't so wild and I rode right into a small herd feeding in a park. There was a big six-point bull elk in the herd. I don't know why I even thought about roping an elk, but I was a pretty good cowboy and I had a really fast horse. I rode right up close to that big bull and threw a loop on him and then things commenced to happening real fast. I dallied him to the saddle horn and when he hit the end of that lariat my cinch busted off my saddle. I hit the ground and he took off dragging my lariat and saddle. I never did find either one.

Trapping Impacts

Other wildlife species that had not been in trouble at the turn of the century were being decreased by human impact by the 1930s. For many years trapping of furbearers and predators was a way of making money. For example, Emma McCrone said,

Charles Dabney and the author unloading hay during elk feeding operations in 1979. Like many ranchers, Charles stocked fish, fed elk and deer and was very intolerant of poachers. The Dabneys operated the San Juan Guest Ranch west of Creede for many years. Photo by author

My mother, Ellen Soward, was one of the first fur traders in the area. She always seemed to have a little money around and the trappers would come to her and sell muskrats, beaver, river otter, mink, weasels, pine marten, bobcats, and coyotes. Our house was two stories next to Trout Creek and I remember sneaking up in the attic with my sister to see rows upon rows of beautiful pelts. When the harvest was over, she would sell the furs in Denver.

Trapping predators and furbearers had become a way of life. It was thought that trapping these animals protected cattle, sheep and other domestic animals. Even though government trappers, forest rangers, and the Game and Fish Department personnel were all trying to exterminate the predators, it was the local rancher and his son who depended on the income gained from trapping. During the Depression years a .22 rifle and a trap line provided the majority of income for some families in the winter months. Trapping was a skill that ranchers and farmers passed down from generation to generation. It was their primary tool to protect their livestock and property from predators and other nuisance animals. For example, George Wintz said that he trapped muskrats because they tunneled through small dams and caused them to wash out. He saw a couple of river otter once and caught the last mink in Bellows Creek. One animal that he saw but didn't trap was the black-footed ferret near Wagon Wheel Gap, because they ate prairie dogs. He believed that when they started poisoning prairie dogs that was the end of the ferrets.

Leroy Brown trapped mink at Brown Lakes:

> I remember a lot of mink when I was a kid. I used to catch them at the ranch at Brown Lakes and oh how they would pile up the fish. In 1917 I remember Dad dug into the lower side of the upper dam next to the spillway and into a muskrat hole. I bet you there was sixty pounds of fish in there. That was a big loss to us. I remember mother yet catching that mink. She just reached down there and grabbed him and choked him to death! I caught the last mink I ever saw when I was taking spawn with "Tuffy" (Allan) Hosselkus at the Santa Maria Reservoir in 1932.

Leroy Brown may have caught the last mink in the Upper Rio Grande, but mink were not eradicated from other drainages in southwest Colorado where they still exist. There were no limits to the catching of predators and furbearers because they had economic value and people gave no thought to their becoming scarce, as most furbearers are nocturnal or live in such dense cover that they were seldom seen and so weren't missed until it was too late. Other species such as beaver, muskrat, bobcat, coyote, pine marten, weasel, badger, red fox, and other small furbearers have had pretty normal population cycles in spite of trapping. Regulations and bag limits began in the 1940s giving some protection to furbearers, while still allowing a sustainable harvest.

Grouse

Some species were at the very edge of their range and were more susceptible to habitat disturbance. Roy Powell remembered seeing mountain sharp-tailed grouse. "When I was young, I would help put up hay on Trout Creek down by the Rio Grande, I remember shooting willow grouse along the river. These were not blue grouse or ptarmigan and they tasted a lot better."

The Forest Service listed the mountain sharp-tailed grouse as present in the Upper Rio Grande. Although not typical sharp-tailed grouse habitat, grouse must have existed in the area of Antelope Park. Elbert Vanaken, Forest Ranger, reported in 1923 that the mountain sharp-tailed grouse were on the increase, but other rangers reported populations near extinction.

The blue grouse of the spruce-fir forests and the ptarmigan of the alpine tundra provided many a dinner to settlers and travelers. Their populations cycled naturally in those areas that were not overgrazed.

Mountain Lion

The San Juan Mountains, Uncompahgre Plateau, and San Luis Valley have a lot of mountain lion habitat. The low lying, brushy, and rocky canyon country in these areas are prime habitat for deer, the major prey of mountain lions. The Upper Rio Grande, because of its lack of deer habitat, is not a prime mountain lion habitat, but they do wander through the region and once in a while one would kill livestock, as reported in the *Creede Candle* in 1925,

> For ten or twelve years a mountain lion has been roaming the range along Rat Creek, and occasionally evidence of its visits are found. A few days ago A.L. Broadhead while riding the range between Rat Creek and Bristol Head peak looking for stray cattle, came across the carcass of a horse that had fallen a victim to the lion. The carcass was literally torn to pieces, large chunks of it being scattered around the vicinity. It was so badly mangled that identification of the brand could not be made.

Ranchers such as George Wintz solved their own lion problems whenever they had the opportunity. Lions came down by the ranch and would kill young colts. They were usually after the deer that lived around the ranch, but would kill whatever they could chase down. Wintz said, "I tracked one lion that had been killing our livestock for ten years at different times. I made up my mind one winter that I would stay on his track until I got him and I did. The killing stopped."

Lions in other areas of southwestern Colorado had more of an impact on limiting wildlife populations, particularly deer. The extermination of wolves in conjunction with increased killing of bears, lions, and coyotes reduced predation as a limiting factor on populations of deer and elk as these populations were beginning to rebound.

Grizzlies

In the 1930s the San Juans had become the last refuge of the mighty grizzly. They were seldom seen, but Roy Powell remembered seeing grizzly in 1933:

> Carroll Wetherill, Don and Dave LaFont, and I were surveying the Weminuche Ditch. This was a ditch that was built to bring water from Rincon La Vaca Creek across the Continental Divide and down Weminuche Creek to the Rio Grande. Carroll would then use the water on his ranch near Del Norte. While we were up there surveying the ditch, I saw three grizzlies in the buck brush over where the North Fork of the Pine River joins the main Pine River. When we first saw them, I told Dave from that distance it looked like a pack horse, because it was so big. He had his binoculars and said it was a grizzly with two cubs following it. She never bothered us.
>
> I've also seen a grizzly up on Trout Creek between the Powell Ranch and the Sulfur Beds mine in about 1930. I've seen tracks where those old claws were out and I knew it was a grizzly, and had seen him a time or two. We didn't realize they were so close to being extinct then. I don't know what happened to it. We never saw it again.

Bears killed cattle, but it was the woolgrower who was most affected by grizzly and black bears. Bears raided bee hives, backcountry cabins, sheep camps, and ranches looking for an easy meal. They seldom bothered people, but one sheepherder had a run-in with a bear on Red Mountain Creek in the summer of 1925 as reported in the *Creede Candle:*

> **Man Seriously Injured by Bear on Red Mountain**
>
> A Mexican sheep herder by the name of Mike Gallegos, who was in charge of a band of sheep belonging to Senator John McFadzean, which had been sent up to the Red Mountain district for summer pasture, was attacked by a huge cinnamon bear Monday afternoon and escaped being killed by the aid of his faithful dog.
>
> Gallegos was trying to find some sheep that had strayed from the bunch and in stepping over a log that lay across his path he landed

Rancher George Wintz, in about 1950, looks after two orphaned fawns on the La Garita Ranch. Photo from Ernest Wilkinson Collection

square on the top of Mr. Bruin who was curled up asleep by the side of the log. The bear jumped up and as Mike explained "It hit me on the side of the head with his left paw." The claws ripped him from just above the temple and stripped the bones clean nearly to his jaw, taking a number of splinters of bone along with it. The bear then came for him with open mouth and was about to seize him by the throat when he threw up his arm for protection and the bear shut down on it breaking one of the bones and severely lacerating the flesh. A little sheep dog which was the constant companion of the herder then took a hand in the affair and made such a violent attack from the rear that the bear let go of the man to deal with the dog and Mike took to his heels and soon put a safe distance between himself and the bear.

Regardless of his serious injuries Gallegos rounded up his sheep and drove them into camp before seeking any aid for his injuries. On Tuesday he was brought to Creede where Dr. McKibbin dressed the wounds after which he was taken on to his home in Monte Vista where he will receive hospital treatment.

> Gallegos is a man of advanced years and the courage and presence of mind displayed in dealing with the bear and his wonderful vitality and power of endurance following the attack are alone accountable for his survival.
>
> The dog whose love for his master, caused him to face death to assist him in a time of peril, was an occupant of the car that took Gallegos to his home and that the bond of affection existing between the man and his pet dog is not doubted by anyone who saw them together on the trip.

The *1934 Annual Game Report* said there were three grizzlies in the Upper Rio Grande. In the 1939 report the estimate was increased to nine. Forest rangers made these population estimates based on reports from sheep herders and were at best just guesses. No one was cognizant of the plight of the grizzly, and it would have been politically incorrect to speak out to save them-after all they weren't all gone yet. In those days there were "good" animals and "bad" animals. All predators were "bad" and one was performing a public service to kill them.

Protecting Wildlife—The Game Warden

In the 1930s people had more important things on their minds than endangered or threatened wildlife. Feeding one's family and surviving the Great Depression was the top priority. There were few jobs and people lived off the land, though not to the extent that the early prospectors did. There were now laws to protect the wildlife, and people were mostly respectful of the attempt to bring the wildlife back. But some people were on the brink of starvation and wildlife was a major source of food. For these people there was one obstacle to living off the land -the Game Warden! In 1930 there were only twenty-four game wardens in Colorado. There were also special deputy game wardens who served on a temporary basis and during hunting seasons.

One of the toughest game wardens in Colorado was Otto Peterson, who served from about 1918 to 1932. Peterson fought for the protection of wildlife with courage and determination, backed up by his physical abilities, and once in a while his pistol. He worked tirelessly from horseback or Model T and made arrests in every county in Colorado. He was a real life "Shadow" who knew when and where to show up to stop illegal killing of fish or wildlife.Clarence Goad was the game warden most people in the Upper Rio Grande knew between

1918 and 1946. He was assigned to patrol the San Luis Valley. During much of this time Bill Krebbs, who lived in Del Norte, was a temporary deputy game warden and he helped Clarence Goad cover the Upper Rio Grande at peak times.

Game warden duty in the early days was mostly law enforcement. A game warden furnished his own vehicle, horse, and other equipment and was paid gasoline, a small expense account, and a salary of about $100 a month. He did receive one-third of the fines that were collected until 1937. He worked alone in wild back country, out of contact with the outside world, and self-reliant for his own survival. There was no officer survival training, only one's instincts. The position and mission of the game warden were highly respected, but wardens were caught between a rock and a hard spot socially. Most people never got personally close to the early-day game wardens, except for their families. Al Birdsey said of those Depression days:

> Everybody poached some. There wasn't any work here. We hung the meat in a cold room at night and wrapped it up and let the air to it during the day time and you could keep a deer and it would never spoil. Now I built a meat house out back to store it when we poached. I kept meat year around, but poached only when we needed meat.
>
> There was one fella who lived out on Moonshine Mesa would go out and kill an elk or two and bring them into his place and dress them. I would go out there with another fella and we'd get them and deliver a quarter at a time to the families. Doc McKibbin was here then and we took him a hindquarter of elk once and he was the most tickled man you ever seen in your life. He wasn't making enough money to eat on and he appreciated it. Meat was always taken care of. If someone would have wasted meat, they would have been run out of town.
>
> We respected the game wardens, but the people went out of their way to see to it that the game warden didn't see anything illegal at all. In fact, my stepdad was a Justice of the Peace for a while. Clarence Goad, the game warden, come up to the house once when my mother had an elk roast in the oven. She let that thing burn up in place of asking Goad to eat with us, because he would have joined us and found it was elk and he would have arrested us. He didn't want to know about any of it.

Clarence Goad was the game warden who worked the entire San Luis Valley and was best known in the Upper Rio Grande between 1918 and 1946. Photo from Division of Wildlife archives

Mabel Wright remembered:

During the Great Depression elk and deer were an important part of many people's diet. Some people didn't have any money to buy meat, but we had cattle and we never cared a lot about the wild meat. Carroll and Gilbert Wetherill, Billy Wills, Howard Kennell, and Floyd Roberts wintered out there at the Wetherill hatchery between San Juan Ranch and Highway 149 in the early 1930s. The mines were shut down. They couldn't get jobs. Kids like that quit school. Not a one finished high school. There was a place to stay and they pooled their resources. They got jobs in the summer. Well, we hired those boys when they were fifteen or sixteen years' old. They hunted and trapped coyotes and would get anything they could from fall to spring. They had to live off the land and wild meat. They used to come to our house five nights a week for supper and come in late in the afternoon and we loved to have them in.

Howard Kennell, reminisced:

Mabel was a really nice lady. We ate many a meal down there. I worked on their hay crew for a couple summers. She fed the whole crew of fourteen to sixteen men. I don't know how she ever did it. We'd be taking eggs in the winter when it was so bitter cold and the phone would ring and she'd say, 'Come down for supper.' She didn't have to ask twice.

Mabel said:

They'd bring us a quarter of elk or a piece of meat and I hate to tell you, but we just couldn't eat it. I tried to cook it. It was tough and wasn't good and we had better meat. I appreciated it because they were trying to help us. But of course they did poach. We laughed and

Mabel Steele Wright in about 1950, displays one of the large rainbow trout caught from the Rio Grande. She cleaned many a cabin and cooked many a meal for her guests. Photo from Charles and Dorothy Steel Collection.

kidded Howard Kennell after he went to work for the Game and Fish Department. He had been a good poacher and knew all the tricks.

It was a rare occurrence for people to see a game warden, because he worked such large areas. But when they did patrol through an area the wardens did catch law violators. Most easily made friends with ranchers, because they were as likely to catch a cattle rustler as they were a poacher. Most ranchers didn't live off the wild game so there was little to fear. Mabel Wright said that Clarence Goad often stayed at their ranch. He patrolled in a roadster and always had his big Irish Setter riding with him.Leroy Brown remembered Bill Krebbs:

> He worked out of Del Norte and drove a car. We lived out at the ranch at Brown Lakes part of the time during those years. During the Depression people lived off of deer and elk. If you got meat and your neighbor didn't have any then you shared it. Game wardens seldom caught anyone. They really didn't try, to tell you the honest truth about it, because there was nobody mistreating it. Everybody was hungry. The only time that they did catch anyone was when a bunch around here was killing elk and taking it to the San Luis Valley and

selling it and they got into trouble. They were warned and then Ranger Fay Franklin and the game warden caught them. We had other meat besides deer and elk. Beef was so cheap. You could buy a whole cow for five bucks.

During the Depression the forest rangers had the authority of game wardens. Leroy Brown's daughter, Janice Nelson described a close call:

> Dad was logging up at Hansen's Sawmill when I was about five years old. One day Fay Franklin, the forest ranger, came up for an inspection. I ran up to him and blurted out, 'My daddy killed an elk and it's buried in the sawdust.' To which Fay asked, 'Are you going to eat it?' and I assured him, 'We won't waste any of it!

Even a child was trained to not waste food. Franklin went about his duties and nothing was ever said. In that day wildlife was there to eat and use, but it was, as we would say today, "not politically correct" to waste it.

Art Davis said of Clarence Goad:

> During the Depression, Clarence Goad was the only Game Warden for the whole San Luis Valley. Goad had to contend with a grapevine that you wouldn't believe. We didn't poach because we wanted to, but we had to eat. We respected the law and the game wardens. There really wasn't as much poaching as you might think. Whenever Goad would leave the Valley to come up to Creede, there would be a moonlight parade of flashlights going up to the trees where people would hide their meat until he went home. He rarely caught anyone . . . except me.
>
> I was working at the Broadhead Ranch and I had bought and butchered a goat. We had eaten most of it, but some was on a platter with some leftovers on the screened porch. Goad came with his warrant when nobody was there and he took it and came down and pinched me and said he was going to take me before the Justice of the Peace for having deer meat, even though I had a bill of sale for the goat. That was in March and it didn't come to trial until the next fall.

Howard Kennell was at the trial:

They had a six-man jury to hear the case. I knew everyone on the jury and each one probably had meat hanging on his back porch. While the trial was going on someone slipped out and got into Goad's rig and stole all the evidence. It was all planned in advance. The defense attorney requested that they show the evidence so the district attorney sent Clarence to get the evidence, but it was all gone and he came back in and the jury started laughing and turned him loose. That is the way it was back in those days.

Howard Kennell had lived close to the land and its wildlife. As with so many of the early trappers and game wardens, some of the best came from the ranks of those who knew by experience the tricks of the trade. Kennell saw a job announcement and applied for a job with the Game and Fish Department in Denver:

I went to work for the state in 1936 as a beaver trapper. Later I became the fur inspector who supervised all the trappers in the San Luis Valley and San Juan Basin. One day I got a call from Goad to help him serve a search warrant on a man I knew who lived down by Wagon Wheel Gap. I met Goad and I said, "Clarence, this man has a wife and a bunch of kids and if we put him in jail, someone is going to have to feed his family." Goad said, "Well, maybe we can figure something out." I asked him where the meat was supposed to be hanging and he said that it was supposed to be on the back porch. So when we got up there I went to the front door and Clarence looked around in back, but after we left he told me he never looked on the back porch . . . "I wanted to give the guy a break." It was only a week later the guy went to town and started telling people that if they ever get caught, let Howard Kennell know and he'll get you off easy. I told them that was the last time. I told Clarence and he said, "That is just the way it is. You get ridiculed if you do and ridiculed if you don't."

" I never let it happen again," said Kennell.

To this day there are families in small western communities that still detest the image of game warden or Game and Fish Department because of conflicts in the Depression years. There were game wardens who had reputations for being heartless, but most of them were family men themselves and were merciful. Punk Cochran the game warden in

South Fork for many years went so far on one occasion to poach a deer and leave it on a front porch of a starving family in the middle of the night. He couldn't tell anyone at the time or he would have lost his job.

The Fishing Camp Era

The wildlife tourism business was beginning to grow during the Depression. The commercial fish business was turning toward supplying trout to resorts. Several ranches such as Workman (Freemon Ranch), Wright's and Wetherill's had built fishermen's cabins as a way of increasing their cash flow in the 1920s. They built small cabins that had a wood burning stove, a table, bed, some shelves, maybe some cabinets and an outhouse. Water was hand-pumped from a well and carried in a bucket. In some cases these cabins have been modernized, but for several generations many families have returned for decades to the precious memories stored in the logs and spartan furnishings of fisherman accommodations.

In several cases husbands died and the widows were left to run the ranch. The Soward Ranch changed from a commercial to a recreational fishery during the Depression. Emma McCrone explained:

> It was in the late 1930s during the Depression and my daughter Margaret and I were running the ranch alone. I would have people come in and ask if they could pay to fish in the lakes. I figured it was a lot easier to let them pay me for fishing than it was for me to set gill nets, and dress fish for market. Well it got evident to me that we were pretty well fixed for fishing. So I made up my mind that people could pay me to fish. I would charge them about $5 a day. Then people would want to stay all night. That led me into the cabin business, because people wanted a place to stay.

Margaret Lamb, Emma's daughter added:

> There was no money. We had sold our cattle. The fences were down. The taxes hadn't been paid. Few know what the Depression was like. People said, "I wish you had a place for us to stay." We got to thinking we could make some money. We had a three-bedroom house and my mother and I would move into one bedroom and rent the other two rooms . . . anything to make a little money. We couldn't wrangle cows and we really couldn't net and ship fish like granddad

did. But, we did have good fishing and we could rent and clean cabins and that is how we got started.

Most resort owners were concerned that their guests should respect the fishing opportunity that they were enjoying. To some guests, being on private land was a license to catch as many fish as they wanted. All the while most guest ranches worked to maintain their excellent fishing by buying and stocking fish and asking guests to stay within the legal bag limits.

Mabel Wright remembered some of the guests at Wright's Ranch:

I remember one time after 1925 that Clarence Goad picked up one of our guests without a fishing license. Clarence was having dinner with us. He was sitting with his back to the door and this woman came in and said to my husband Ray, "I really got myself in trouble", and she started laying this whole thing out. Then she told Ray, "You gotta get me out of this." Nothing bothered Ray, "Well, this is Mr. Goad and he can tell you what you need to do." The next day they went into Creede for court. I thought it was so funny.

I don't remember that there were a lot of violations. My goodness what could anyone do with more than a limit of twenty fish? Most of our guests weren't fish hogs. Wallace Wright was very interested in stocking the river and he kept warning people not to take more than their limit and I wish people would have listened. You know we never took fish from the state, we didn't want theirs, but we bought fish and put in the river.

We bought any kind of fish we could get, but in those days we could get natives and then rainbow and brook. I don't remember browns until the 1930s or 1940s. When I first came, Father Wright would bring cans of brooks and rainbows in an automobile. He was always encouraging the neighbors, "Get some fish, get some fish." He offered fish to anyone if they would just return the cans. Nobody was interested. Everybody would say, "Oh there will always be a lot of fish." You even heard that from old timers. People thought there was an endless supply of fish. Workmans for instance had a little resort business and they said, "Oh no." I remember Wallace getting unhappy when somebody would come in with too many fish. They didn't have to be over the limit, but he would ask what were they going to do with

Fishermen holding a stringer of more than seventy-five trout at Ruby Lake. (circa 1940). Photo from Kipp Family Collection

all those? Well people ate fish and used them and I really doubt they wasted much. I think that they really did get them because they wanted the meat. The ones that were really bad about taking too many fish were from Pueblo and Walsenburg. They camped out in places like Box Canyon and they would catch too many fish and Wallace would turn them in because they were taking more than their share and they were depleting the stream. They salted them and brought kegs and containers to preserve them and eat them in the winter. But I remember Wallace turning in a guest of ours from the San Luis Valley. This man had been fishing over on Clear Creek and he brought back way too many and he had a whole dish pan full of beautiful trout twenty inches or more. He said, "I sure slaughtered them today." My husband Ray was so indignant and said, "You'd think he was some kind of predatory animal that should be praised for getting rid of them." You can't stock water fast enough for people who fish like that.

Values and beliefs of that time brought forth traditions. It was socially acceptable to stock fish, kill predators, feed deer and elk in the winter, poach out of season as long as the meat wasn't wasted, and other traditions. Many of those traditions have been handed down by families and community and were accepted as right. New ideas and technology were resisted and change came slowly, but it did come.

Chapter 6

WILDLIFE ALSO SURVIVES THE DEPRESSION AND WAR YEARS

With few exceptions most wildlife species in Colorado increased through the Great Depression and World War II years. Wildlife management based on science started during this time as research biologists sought answers to basic questions about wildlife biology and population status of different species. The word "ecology" was added to the wildlife scientists' vocabulary as they began to learn the interrelationships of water, soil, plants, wildlife, and people. The management of wildlife resources would soon be based on this knowledge and modified by the democratic process. As big game populations continued to increase conflicts between agriculture and wildlife multiplied. Government programs were dedicated to the elimination of all predators that threatened livestock. The legislature required the Game and Fish Department to pay ranchers for the damage done to crops and haystacks by wildlife.

Politics continued to play an important role, but the Colorado Legislature created a citizen Game and Fish Commission to direct the Game and Fish Department in 1937. Advancements were made to improve fish stocking. The role of the game warden was expanded beyond wildlife law enforcement to a district wildlife officer who was responsible for not only law enforcement but also improving relationships between sportsmen and landowners and find places and the means to increase wildlife numbers. With all hunting and fishing license money being deposited in a Game Cash Fund, the Game and Fish Department was able to expand programs to purchase lands and waters that were managed for wildlife and provided public areas for hunting and fishing. These were exciting times for wildlife and sportsmen in Colorado.

Conflicts Between Agriculture and Wildlife

By the early 1930s Colorado's deer and elk herds had increased to the point that ranchers were complaining that they were causing damage to their private property. Their outcry was so loud that in 1931 the Colorado Legislature passed the first game damage law to require the Game and Fish Department to pay ranchers for damage done to crops (mainly haystacks) and for livestock killed by bears and mountain lions. These payments were to be paid from hunting and fishing license money in the Game Cash Fund. The ranchers also demanded that elk herds be reduced. The Legislature responded by opening the first elk season since 1903 in eleven Colorado counties, but only Hinsdale and La Plata counties in the San Juan Basin area were open in southwestern Colorado. Elk hunting would not resume in the Upper Rio Grande until 1938.

By 1934 the *Annual Game Report* noted that there had previously been no elk north of the Rio Grande, but they were increasing and becoming common throughout the Upper Rio Grande. The report listed about 900 elk, 150 bighorn, 1,400 deer, 95 black bear and 3 grizzly in the Upper Rio Grande. Forest Rangers made these estimates. Forest Service personnel reported that elk sightings and tracks provided evidence of widespread migration of animals all over the San Juans.

In 1935 the State Board of Livestock Inspection entered into agreements with the Bureau of Biological Survey and the Department of Agriculture to exterminate predatory animals. Woolgrowers paid a tax for each sheep they owned into the newly created Predatory Animal Fund. These assessments continued into the 1950s. In 1941 the Game and Fish Department turned its predator control efforts over to the U.S. Fish and Wildlife Service, paying the expenses of government trappers from the Game Cash Fund.

Colorado Game and Fish Commission

Until 1937 the Colorado Legislature set hunting and fishing regulations, which were based on political pressure from ranchers, farmers, sportsmen and business interests. In 1937 Colorado legislators decided that they should not be involved in the day-to-day operation of the Game and Fish Department, so it created the Colorado Game and Fish Commission, which was composed of six citizens appointed by the governor and approved by the Senate. The Commission was given the authority to establish hunting and fishing seasons, set bag limits, determine legal means of taking wildlife, and provide policy direction to the Game and Fish Department. All revenue from hunting and fishing licenses was to be kept in the state treasury for the exclusive use of the Game and Fish Department. No general tax revenue was to be used to support wildlife protection in Colorado. The legislature did reserve the authority to approve the department's budget, to set license fees, and to create wildlife statutes. Otis E. McIntyre was appointed to represent the San Luis Valley on the first Game and Fish Commission.

In their first meeting the commission addressed the issues of financial accountability, fish hatchery accountability, approval of more stream-side pond agreements, a state trapper's service, uniforms for game wardens, bait and fly-fishing restrictions, hunter safety, and law enforcement. The commission started a publication, *Colorado Conservation Comments,* a report to all sportsmen and other interests, which became *Colorado Outdoors* magazine.

Elk Seasons

Although elk hunting season had opened as early as 1931 in some Colorado counties, the Game and Fish Commission opened the first elk season in Mineral County in 1938. There was little fanfare in the press. Many local hunters went hunting and killed a bull. One didn't have to venture too far into the back country to find a good bull. The elk weren't very wild then, but it didn't take them long to learn to avoid humans. Most hunters were local men, because the Depression kept many people from traveling long distances for such pursuits as hunting and fishing. Howard Kennell and Roy Powell had a typical elk hunt together in that first season:

> Roy Powell and I rode horses a short distance up to where Trout Creek and Jumper Creek come together and camped. The first morning I shot a spike bull. Roy caught about eight head crossing a snowslide area on Jumper Creek and he shot a big six-point bull. Hunting elk was pretty easy, even though there weren't as many elk as there are now, but the chore of packing that big bull off the mountain was hard work.

There was a fever of excitement that started in September as hunters began to prepare and dream of the hunt. After the hunting season was over in October the local newspapers were filled with the names of successful hunters and great status was given to those who had killed the largest animal. Elk seasons became an annual event by 1939 in most mountain counties of Colorado, but the elk hunting season almost didn't happen in the Upper Rio Grande that year. It was the Game and Fish Commission policy at that time to allow elk hunting only when a county commission requested a hunting season in their county. The Mineral County Commissioners had not filed a request for a hunting season in time for the Game and Fish Commission to consider their request. As isolated as they are from the seat of state government, most western Colorado communities either got involved in wildlife management decisions and projects or they were ignored as reported in the September 20, 1939, *Del Norte Prospector*:

> **Mineral Added to List of Elk Hunting Areas**
> . . . Mineral and Saguache counties will have an open elk season this year . . . These two counties, originally scheduled not to have the elk season, were added at the last minute . . . Mineral County residents had petitioned the State Game Commission six weeks ago for the open season, but were informed that the request came in too late . . . The three county area of Saguache, Mineral, and Rio Grande had open season on bull elk last year . . . The only other change in the big game season made by the Game Commission was declaring that the bow and arrow will be legal weapons for hunting deer, elk and bear this year. Many archers and devotees of the sport of Robin Hood and Little John had requested permission to hunt with bow and arrow. Bows must be at least 50 pounds weight and arrows must have keen broad heads that are at least an inch in width . . .

Since there were few hunters and the first few years of hunting were for bulls only, the elk continued to multiply. The commission continued to set seasons and bag limits based on the opinions of politicians, businessmen, and sportsmen. The Game and Fish Department didn't have biologists nor any systematic way to take a census of deer, elk or any other wildlife. The only field personnel were game wardens who were law enforcement oriented, even though they were very knowledgeable about wildlife populations. The Forest Service had been estimating numbers of deer and elk for several years, but they weren't used either. There was a need to base wildlife management decisions on a dependable census, but wildlife science and the art of management had not yet been born.

Wildlife Population Inventories

In 1937 the U.S. Congress passed the Pittman-Robertson bill, levying an eleven percent excise tax on firearms and ammunition, that was earmarked for wildlife habitat restoration and research. This money was allocated by a formula back to the states. Colorado began participating in this program in 1939 by matching license revenue with the federal funds. Funding was used for a count of deer and elk, and for other research projects. The funds also purchased critical wildlife habitats throughout Colorado such as the Coller State Wildlife Area near South Fork.

The Game and Fish Department began to hire biologists to increase the scientific knowledge necessary to manage Colorado's game animals and birds. One of the greatest management needs was for a better inventory of deer and elk. There was no statistical way of determining how many big game animals there were. Norman Kramer, a mountain pilot from Alamosa, flew the first airplane counts of deer and elk in 1938. Kramer developed techniques for finding and counting animals from the air and he trained additional pilots and observers. The information from these counts began to form a basis for establishing seasons and bag limits.

Several wildlife species were creating problems that in time became opportunities to increase knowledge. A beaver census was just the beginning of the Game and Fish Department's effort to manage a wildlife species based on a population estimate. In the late 1930s beaver were becoming too abundant and causing problems for landowners. Since the State of Colorado protected beaver, the legislature made it the responsibility of the Game and Fish Department to manage them. The legislature passed the 1941 Beaver Control Act that directed the department to respond to the demands of ranchers and farmers within twenty-four hours and to trap all nuisance beaver. The Department hired a corps of beaver trappers to meet this legislated mandate. Howard Kennell became the first fur inspector who supervised the fur trappers in the San Luis Valley and San Juan Basin:

> One of the first things headquarters wanted us to do was conduct a beaver census so we would know how many beaver we had and how

Howard Kennell in about 1940 with a small beaver that he used in school programs to teach children about beaver. Kennell was the first fur inspector to supervise trappers in the San Juan and Rio Grande Basins and retired as the Area Supervisor in the Gunnison Basin. Photo from Howard Kennell Collection

many should be trapped each year. I hired Allison Mason, a good trapper over in Durango, to help me out with a beaver survey on the Pine River toward the Continental Divide. We had to count beaver dams and food caches and estimate a beaver population for each drainage. We spent a lot of time riding, fishing, and camping in that beautiful country. We did this kind of work in all drainages throughout the San Juan Basin and the San Luis Valley.

I hired Glenn Cochran, Punk Cochran, Billy Schultz and Carl Welch to trap beaver in the San Luis Valley. Then World War II came along. During the war Wayne Nash was the fur inspector in the San Luis Valley. In those days we trapped and skinned the beaver, but when we took them on private land, the landowner got half of the price of the pelts, which were bringing in from $33 to $57. The ranchers and farmers really wanted us to trap beaver, because it was free money without any work on their part. We also live-trapped and moved a lot of beaver. We packed many a beaver horseback into the San Juan back country to put them on streams where they would stay and build dams. Soon the beaver were increasing too fast and they were getting out of control, flooding fields, and roads, and culverts, and blocking trails, and ditches, and canals, and cutting down desirable trees.

Bill Schultz was a typical example of how many men started careers in the Game and Fish Department following World War II:

I started trapping beaver under Howard Kennell in 1941 and a year later Howard joined the SeaBees in the Army and I went to work under Wayne Nash and I worked for Wayne until men in the trapping division began joining the warden service. Glenn Cochran, Carol

Welch, Wayne Nash, and I trapped 1,200 beaver in one spring in the Conejos, Saguache, and the Upper Rio Grande. We were just taking the beaver out of irrigation ditches and the ones that were bothering in the high country plugging up culverts and the like. We didn't try to eliminate them.

They paid my gas, three and a half cents a mile, and $125 a month. I had to furnish my vehicle and equipment. That was a pretty good job in those days. So the first couple of years all I did was trap beaver, except when they were having a raid or something like one time we went to Bonanza with Carl Welch, who was also trapping beaver at that time in the southern end of the San Luis Valley. The Warden Service came in and brought in all the trappers and we seized twenty-six illegal deer in two days. They were fined about $100 a deer.

I transferred from being a Trapper to Game Warden when I was in Saguache. I transferred to Creede in about 1946. There were other wardens: Carl Welch was in Antonito, and Dick McDonald in Monte Vista, then in a year and a half they moved Glen Holzworth in here. Later Punk Cochran went into Del Norte.

Howard Kennell hired Martin Burget to trap beaver in Montezuma County area before the war. Burget was a good trapper, but when most of the trappers joined the game warden service, he became the state turkey trapper and known as "Turkey Burget." Colorado had seen tremendous die-offs of turkeys because of disease and overkill during the Depression years. The only populations of turkey were found near Trinidad, Pagosa Springs, and Durango. The department used Pittman-Robertson funds to reestablish turkeys in other areas of Colorado. The department had experimented with domestically raised turkeys, but those attempts had ended in failure. In 1942 the department bought property on Devil Creek west of Pagosa Springs and began an intensive turkey restoration project. The property was developed into a site to attract wild turkeys so they could be caught and transplanted. As a result these wild birds were then transported to other locations throughout Colorado.

The Impact of the War Years

By 1940 deer populations were beginning to soar and elk were right behind them. Years of protection had brought them back from near extinction and in many areas had developed protectionist attitudes toward these animals. People were fearful that hunting would wipe them out again. Throughout Colorado the deer population was increasing beyond the winter range carrying capacity, and some major winterkills had occurred because of starvation. There had been many feeding operations that were mostly failures, because they tended to concentrate deer, which increased mortality because of disease. The deer could not digest hay and died from the feed itself. The department felt it was better to reduce the deer herds to the carrying capacity of their natural ranges by harvesting some of the doe deer. The commission established the first doe permits since 1907 in order to reduce some deer herds, and there was a public outcry. Liberal seasons and allowing doe deer to be taken by hunters were very unpopular in most western Colorado communities.

In 1946 a bear could be taken on a deer or elk license during the deer and elk seasons. The license was good for either black bear or grizzly although there were only an estimated nine grizzlies in the San Juans and since 1940 those had been in the Rio Grande National Forest.

Cleland N. Feast became director of the Game and Fish Department in 1940. Feast held several positions prior to becoming director, one of which brought him to southwestern Colorado. Feast applied for a school teaching job in Creede and was Superintendent of Creede Schools in 1931. His daughter Carolyn Feast McCracken said:

> Dad and Mom loved Creede and the people who lived there. They lived in a single family residence that used to be a saloon right next to Willow Creek. Mom said it had wood flooring with huge cracks showing the dirt underneath, no water or plumbing, and no electricity. The water was brought into the house from the creek. In the winter she used an axe and brought in the ice blocks and melted them on a potbellied stove. She said the winters were cold but the closeness of the entire community helped each other survive.

Feast left Creede and spent the rest of the Depression years in Colorado Springs managing the Broadmoor Hotel. Soon after he became director World War II started and this caused a real dilemma for him. Feast had to balance war rationing and wildlife conservation. Since field people were in the back country, he had all law enforce-

By 1949 deer were beginning to overpopulate some winter ranges in Colorado, setting the stage for a major die-off in the Gunnison Valley. Photo from Laurence. E. Riordan Collection

ment and trapping personnel trained for civilian defense and sabotage detection. Feast appointed 500 special wardens to serve without pay during the war. World War II turned most attention away from wildlife management, research, fish stocking and record keeping. Wildlife continued to be a major food source for many local people, but not as many people came to Colorado during those war years. Gasoline and tires were rationed, yet a few people still had a way of saving up their ration stamps and taking at least a fishing vacation.

Using Pittman-Robertson wildlife research funds after the War, the Game and Fish Department considered the renewal of a deer and elk inventory as a high priority. In 1948 C.D. "Con" Tolman flew the first aerial count for elk in the Gunnison and Rio Grande drainages and counted about 1,000 animals. Ground counts were also made and management decisions began to be based on information that had some scientific basis rather than just anecdotal observations.

The deer population throughout most of western Colorado was growing beyond the winter range carrying capacity. The winter of

1948-49 was very severe in the Gunnison valley, and there was a large die-off of deer. Wildlife professionals feared this situation could be repeated in other winter ranges throughout Colorado. Laurence E. Riordan, former Deputy Director, said:

> I made a reconnaissance trip with Game Manager Gil Hunter, Assistant Director John Hart, and C.D. Tolman into the Gunnison area after the severe winter of 1948-49. I remember one instance when we found six dead and dying deer under one juniper tree in the area around Elk Creek where Blue Mesa Reservoir is now. Our problem then was not from failing to recognize the situation of overpopulations of deer, but was a combination of ignorance on the part of the general public and a lack of hunting pressure to accomplish the needed herd reductions. We were fortunate to have the legal authority and support of the commission to set liberal deer seasons, but there were just not enough deer hunters at that time in Colorado to accomplish the task.

Bill Schultz said of that winter in Gunnison:

> I remember in the spring of 1949 they took every game warden and trapper in the state and we went over to Gunnison in the spring and we buried 5,800 deer carcasses that had died on feed grounds and along the highways. We only moved those deer that were in the public eye. Hay doesn't work well for deer so they fed deer willow pellets made in California and they still died. But, when you have a year like that the Game Department said it was a lot better to allow the public to take those deer than it would be to bury them each spring. We had two bulldozers over there digging trenches. The guys were hauling deer with three hay trucks. There would be five men on those hay trucks and two men in each pickup hauling deer. We were afraid we might start seeing the same thing on the Rio Grande. The deer in the lower areas were becoming too abundant for the feed on the winter range. Ranchers were complaining about getting lots of game damage. We were just trying to put more hunting pressure on the deer. In some areas there were just too many deer. If we could have had a special season in certain areas then we wouldn't have had the problem with too many deer for the range.

In spite of a major deer feeding operation in the Gunnison Valley in 1949 over 5,000 deer died on feed grounds and along highways. Game and Fish Department game wardens and trappers collected, buried, and burned the carcasses. Photo from Laurence E. Riordan Collection

The state hired two deer herders, but they couldn't keep up with the demand so we all worked to prevent game damage. We were pressured by ranchers and farmers to protect their crops and I spent many a night in South Fork chasing deer out of the fields where they raised peas and lettuce and tried my best to protect the farmers' crops.

We made some mistakes in hunting some deer herds too low. The state allowed multiple deer licenses that allowed a hunter to take up to four deer. They went too far. We also had to fight for wildlife with the Forest Service. They wanted even more deer to be killed. I heard a Forest Service Supervisor get up in a meeting in Monte Vista and say, "Let's kill every damn deer and elk in the country and then we won't have any trouble." There was a time when the Forest Service didn't want a deer or elk on the forest. There has been a change in attitude by the Forest Service as well as the Game and Fish Department. It was quite a fight to cut a deer hunting season to bucks only. The Forest Service would fight us on it and say that the deer were already eating too much grass, but they wouldn't reduce the sheep and cattle on the allotments.

Elk were not as abundant as deer after World War, but they too were increasing and in another decade there would be more elk than deer in the Upper Rio Grande. This was the reverse of trends in the rest of southwestern Colorado where the habitat favored larger deer herds.

The state used multiple licenses as a lure to increase hunting pressure in the late 1950s and early 1960s to reduce the deer population in most of western Colorado. The hunting pressure was more than adequate to accomplish the task. But deer hunting seasons have had little to do with the ups and downs of the deer population in the Upper Rio Grande. The winter range in the Upper Rio Grande is unique in Colorado. There is only grass and no quality shrubs. Deer are primarily browsing animals and in most wintering ranges of Colorado the largest herds are found where there is a mixture of sage brush, mountain mahogany, bitter brush, and other palatable and nutritious shrubs. Such ranges are found in the San Juan Basin, Gunnison, and Uncompahgre and in most of northwest Colorado. About the only shrubs in the Upper Rio Grande which deer eat are skunk bush, gooseberry, currant, snowberry, and forbs which have low palatability and nutritional value. Needless to say the Rio Grande deer don't have all this information and really do quite well in mild and average winters. Severe winters, however, are devastating to deer living in such a marginal habitat. There is better deer habitat in the foothill country surrounding the San Luis Valley, because of the increased variety and abundance of browse.

Bighorn sheep

During the first half of the twentieth century bighorn sheep had recovered from their turn- of-the-century population levels, but in the early 1950s they were again crashing toward extinction. Throughout Colorado bighorn sheep were dying off from pneumonia caused by lungworm. Little was known at that time about the life cycle of this parasite, but it was known that bighorn stayed in the same habitat and didn't move around. It was also thought by some that inbreeding could be a problem. In 1953 the state opened the bighorn season in just a few areas in Colorado, and only a few licenses were issued. Ten bighorn sheep permits on Pole Mountain were allowed. That population hadn't crashed yet, but it was feared that it would be like other herds in the state. The Game and Fish Department biologists hoped that a limited hunting season would scare the bighorns from some of their traditional bedding grounds and perhaps take some of the old rams. This was the first bighorn season in sixty-eight years. The Department wanted to ensure that only the older rams would be taken. A game warden was assigned to each hunting area, and the hunters had to camp adjacent to his camp and hunt under his supervision. The hunt had little impact on the Pole Mountain herd. Some animals were taken, but the herd eventually dwindled to less than twenty head and has remained at that population level ever since. Most bighorn herds throughout Colorado crashed during the 1950s, which stimulated new research that produced new knowledge about bighorn diseases and parasites that was applied during the 1970s to help bighorns recover in many of their historic ranges.

Waterfowl Research

More wildlife research was also being done all across Colorado in the late 1940s. Pittman-Robertson money was provided through the Colorado Wildlife Cooperative Unit at what was then Colorado Agriculture and Mechanical College (Colorado State University). Many wildlife research projects were conducted to fulfill requirements for master and doctorate degrees in the early 1950s. Dr. Ronald Ryder inventoried the waterfowl and wetlands in the San Luis Valley to determine the most productive nesting areas. Over the years some of those priority habitats have been purchased by private, state, and federal agencies and have became waterfowl refuges. Careful management of these properties created what led to the San Luis Valley being nicknamed, the "Duck Factory of Colorado."

Hunters Take Action

Although deer and elk populations were increasing, the hunting pressure after World War II dramatically increased, and there was some fear that hunting could once again destroy all the progress that had been made in recent decades. Sportsmen in most western communities reorganized established rod and gun clubs to support wildlife management. In 1950 the San Luis Valley Rod and Gun Club at Monte Vista as well as the Del Monte Gun Club organized "to conserve and restore the fish and wildlife resources of the valley, cooperate with existing agencies, work for better hunting and fishing in the valley, and develop better relations between land owners and sportsmen."

After World War II there was a tremendous increase in the number of hunters and anglers. Hunting and fishing was becoming poor and such an event was bad for outfitters and other sportsmen dependent businesses. Those who provided services also began to organize to address their needs. In 1947 Bibs Wyley of Cottonwood Cove, Glenn Cochran of Sky High Ranch, and Glen Caron of Riverside Ranch called a meeting to organize the Upper Rio Grande Fish and Game Association. Just as outfitters, resort owners, and business people had organized in many places in Colorado, this local association assisted the Game and Fish Department with stocking fish and provided a forum for them to have a greater voice in wildlife management decisions. The Upper Rio Grande Fish and Game Association expressed its concern as reported in the April 17, 1953, *Del Norte Prospector:*

> **CREEDE SPORTS GROUP FAVORS DEER, ELK BAN**
>
> Say Herds in Danger by Heavy Hunter Traffic
>
> Creede-The Upper Rio Grande Fish and Game Association voted on Sunday to recommend a closed elk and deer hunting season on the Rio Grande river watershed for the coming year. Meeting here Sunday a lengthy discussion was conducted on the big game situation. Sportsmen pointed out that unless hunting is stopped on the dwindling herd, the deer and elk will soon be at a point where it will take 25 to 30 years to build up the game herds again to normal. The unanimous vote on the recommended closing to be sent to the State Game and Fish Commission followed the comments. The Upper Rio Grande region has been one of the heaviest hunted sections in the state in recent years, since the booming big game business. Both resident and out-of-state hunters have swelled. Last fall there was a noticeable drop in kill and poorer hunting conditions.

It was becoming obvious that the elk population could not sustain unregulated hunting. Citizens at that time were familiar with two approaches to managing wildlife: unrestricted hunting and totally closed seasons. Hilda Kipp, former owner of the Wetherill Ranch, said that the concern for the dwindling elk herd had become so great that a number of people in the association were considering a proposal by a trucking company which offered free transportation of elk from Yellowstone if the locals would raise five dollars a head to bring elk back to the Rio Grande. The Forest Service and Game and Fish Department disapproved of the proposal, and it didn't gain public support. Hilda said, "We could have gone ahead and brought the elk in and dumped them out on our ranches. We couldn't have helped it if they could jump fences and get away onto the forest."

Rather than yield to "put and take" elk management the Game and Fish Department recommended that some control on the killing of cow elk be started. The Game and Fish Commission responded by instituting the first cow elk validation system. The validation was merely an ink stamp on a license declaring it valid for cow elk. Validations were allotted by game management units. For example, the commission allowed one hundred cow elk validations for all of the Rio Grande drainage west of Monte Vista. The commission reserved fifteen percent of the validations for nonresidents. It was quite a process to get a validation. A person bought a bull elk license from a license agent and then at the appointed time lined up at selected sites in each county. The first one hundred license holders would have their licenses stamped.

Since the legislature had required the Game and Fish Department to pay for game damage to ranchers, the herds of deer and elk had increased dramatically, and more and more license money was going to complaining ranchers. After World War II sportsmen were getting the feeling that the Game and Fish Department was becoming an agency that was just funneling their hunting and fishing license money to ranchers. To add insult to injury, sportsmen were being denied previously available access across private lands to public lands and fishing areas, even though ranchers often had legitimate reasons and the legal right to close their lands to the public. Some of these closures, however, were illegal. The Game and Fish Department could not use hunting as a means of controlling the deer and elk populations on private lands if access was denied. Rather than funnel away sportsmen's money to ranchers for game damage the Game and Fish Commission established extended seasons, post seasons, and multiple deer licenses to increase hunting pressure and generally reduce the deer populations on public lands. It appeared to the public that the department was just selling off the deer and elk herds to make itself rich. But the reality of the situation was that "biological carrying capacity" of rangelands changed to "political carrying capacity" defined by rancher and farmer tolerance for deer and elk on their lands as well as on public lands where they had grazing permits.

During this time the livestock "barons," which included not only ranchers but also Forest Service career officials and politicians, were committed (as evidenced by their policy and management decisions) to grazing domestic livestock to the detriment of wildlife. However to stereotype all ranchers as anti-wildlife is grossly unfair and inaccurate. For example, the Upper Rio Grande ranchers have been far more tolerant and protective of its wildlife than those in many western communities. For example Dan Soward and Frank Coller were ranchers who took the lead to bring back elk but at the same time deplored having deer wintering on their ranches. A few wild animals were tolerable and added excitement to rural living; but being descended on by large numbers of deer or elk became too much of a burden for most ranchers, who needed the feed for their livestock.

Mabel Wright remembered elk causing damage to their haystacks:

> The Wrights began having real problems in the winter of 1952. It was a hard winter and started early with deep snow and we had elk problems from then on. The Game and Fish Department paneled our haystacks. We used to winter 500 head of cattle plus our yearlings. The only problem we had was the elk getting into our haystacks. Before we had the stacks paneled, I used to go with my husband and we would try spooking the elk, but we always felt sorry for them. It wasn't what the elk ate, but what they destroyed by urinating, defecating, and trampling that was the problem. They got on top of the hay stack and they would play and I think they ate only the choice morsels. The elk waited until we started hauling hay in the evening and Ray would take his rifle and shoot over them to scare them away. But they would come back as soon as we left to feed the cattle.
>
> Not as many animals died as you might think. You know it was after that there were so many that died, because the Game and Fish

> Department had to panel the stacks so there wasn't any hay for the elk. We never found any dead animals around our haystack, except for a calf that got tangled in the fence panels. We never had enough deer around here to even think about them.

In the mid-1950s several ranches in the Upper Rio Grande sold off their cattle and leased their meadows to other cattlemen who grazed their animals all summer and then shipped their cattle to the San Luis Valley in the fall. Only enough hay was put up to feed horses. Still other ranchers continued the management practices established during the homesteading era that connected the livestock operations of their ranches to grazing on public land. These cattle operations put up hay on their summer pasture while their cattle graze the high country public lands. Then each fall the cattle are rounded up and herded to the home ranch where the base herd is fed native hay all winter and the surplus is taken to market.

Managing Bears

For many years the only attention given to bears was for their predation on livestock. Government trappers caught bears, and a hunter could kill one without a license. Through the 1950s, however, there were bear seasons that coincided with the regular deer and elk seasons. There wasn't a bear license, but one could be taken by a hunter who had either a deer or elk license, and the bag limit was two bears. A special spring and summer bear season was established with the hopes of reducing the bear population to control the game damage that bear might cause to domestic livestock. Sightings of grizzlies were very rare, but sheep herders occasionally reported them. Ernest Wilkinson was a government trapper working in the San Luis Valley and he said:

> I killed one of the last grizzlies on the Rio Grande up in Starvation Gulch in 1951. It was killing sheep. I didn't know it was a grizzly until I saw it dead. That same year another one was taken by a sheep herder over at Platoro, Colorado. I think it was a young grizzly and when I came down in the fall Wiley, of the Skyline Lodge, gave me the hide. It was very poor, but I mounted it and it is still down at the Skyline Lodge in Platoro.

Starvation Gulch above Brewster Park where Ernest Wilkinson killed the last grizzly in the Upper Rio Grande in 1951. The bear was killed near timberline. Photo by author

I learned to trap as a boy. What brought it on was that my dad became paralyzed and there was nobody to feed our stock, so I quit school in the tenth grade so I could take care of the livestock, and during the winter after I had the cattle fed and all the chores done, I would trap here and there and pretty soon neighbors would call about a badger in the hen house or something and I got a reputation of being a good trapper. Lee Bacus, a government agent, asked if I wanted a job trapping bear and I said that I didn't know anything about bear and he said that was okay. So he sent a man up with me for three days to teach me how to trap bear and deal with the sheep herders.

I caught my first bear at Goose Lake. I knew some trappers who whenever they saw a bear they took it because they could take the ears and got credit for it. I never killed a bear that wasn't doing damage. If I was going down the trail and saw a bear right there in the open, I wouldn't shoot it. There was a time or two that there were bear near sheep and several weeks later I would get a call that a bear was killing sheep and I would have to go in and take it.

When you had one that was killing sheep you were more or less trying to take the one that is doing the killing. You had a carcass there and lots of times they would be moving through and if you go up and set a trap, they will never come again. But, other times they had a habit of killing two in one night. They would eat the one and then come back the second night and eat on the other one and the third night they would come back and kill more. I learned to roll out of my sleeping bag before daylight and in the mountains there was always dew on the grass and I looked for the closest finger of timber and look for where the bear had come out of the timber onto the grass. Then I would take my lariat and drag a carcass to that point of timber. Then I would build a triangular pen with a top on it so the bear couldn't come in from the top. I'd put the carcass in the back of the pen and set the trap so the bear would have to step on it in order to get to the carcass. At that time we were required to cut the ears off and send them in to verify the kill.

During the Depression, trapping is what got some families through the winter. As far as ranchers, if you don't mind me saying, they got spoiled. They would hear a coyote howling and they would call a trapper and those coyotes weren't hurting anything. Then other times I would go up on a call where a rancher reported bear killing cattle and I've seen where lightning killed a cow and the rancher was blaming bear, because a bear came by to scavenge the carcass. The bear was being blamed and the same with sheep, when coyotes killed sheep. All the rancher wanted was to get paid by the Game Department for a bear killing the animal.

When the government started using poison, I quit. I couldn't see the use of poison, because it killed everything. I was known to be a lone wolf and that was a compliment. During the Depression I had a lot of animals that I couldn't afford to have mounted so I took a correspondence course and started mounting animals for the neighbors and after I had finished up working in the fall then all winter I would do taxidermy until summer. It was more of a hobby and when I quit trapping I started doing it full time. I can't stand being indoors all the time so I started doing tours and now I am teaching primitive skills to the Southern Ute Indian children.

Ernest Wilkinson, wildlife trapper, photographer, taxidermist, outdoorsman, and educator inspects a sheep carcass killed by coyotes in the Upper Rio Grande. (circa 1949). Photo from Ernest Wilkinson Collection

Lloyd Anderson, a government trapper in the San Juan Basin, killed what was considered the "last" grizzly in Colorado near the headwaters of the Pine River south of Rio Grande Reservoir in 1952. There have been no verified sightings since then, although Anderson said that he had seen grizzly as late as 1967.

The Game and Fish Department initiated its first search for grizzly in 1955. That summer Mitchell G. "Red" Sheldon was hired to ride the newly established San Juan-Rio Grande Grizzly Bear Management area and search for grizzlies. The commission prohibited the hunting of all bears in that area from 1955 to 1964. The grizzly refuge was located south of Rio Grande Reservoir (between Squaw Creek and Beartown) and on the Vallecito, Pine, and Williams Creek drainages on the San Juan side of the Continental Divide. At the time this was thought to be the last grizzly country left in Colorado.

Sheldon rode all summer looking for grizzly and talking to sheep herders. Sheldon wrote the story about his search in the January, 1956,

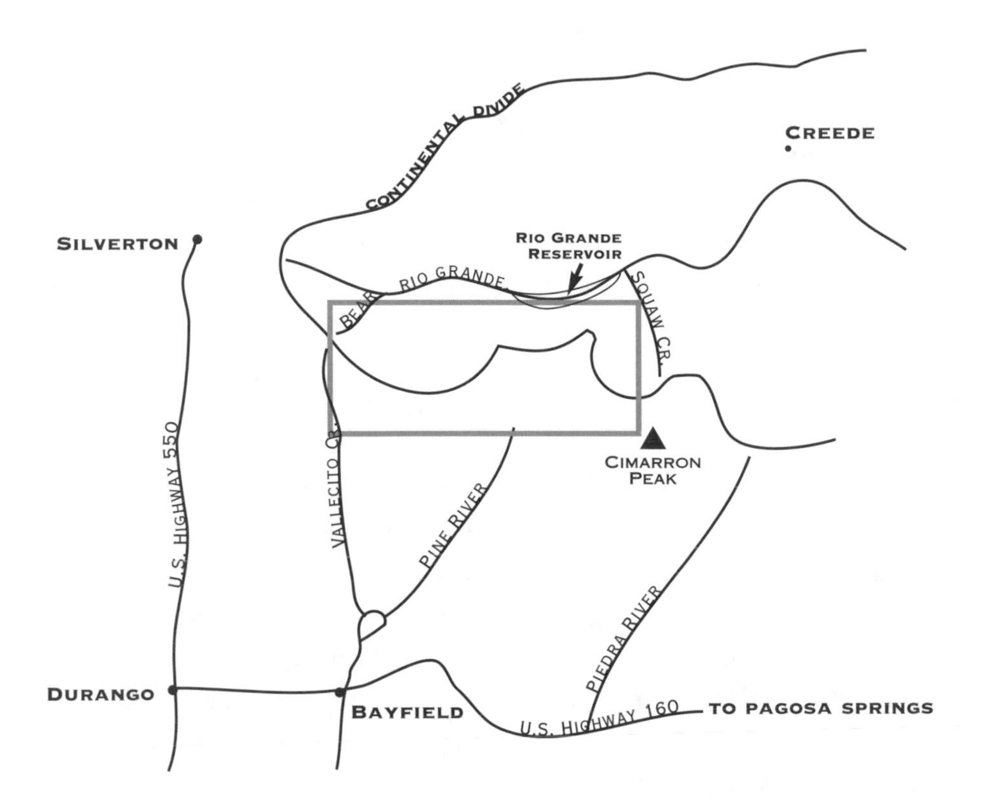

San Juan–Rio Grande Grizzly Bear Management Area. The last grizzlies killed and seen in Colorado were mostly within the boundary of the San Juan-Rio Grande Grizzly Bear Management Area.

issue of Colorado Outdoors Magazine and said:

> . . . Well, I just rode and looked. I had a base camp in a cabin pretty well up on the Rio Grande side. I did my best to scour the territory with a horse, or when I couldn't ride, on foot. I looked for anything I could prove was grizzly sign . . . All told I learned of four grizzlies that had been sighted by sheep herders during the summer of 1956 . . . By the time I came off the mountain I figured I'd traveled a thousand miles on horseback or afoot. I hadn't found a grizzly.

It would take twenty-seven years from the time Anderson killed his last grizzly until the existence of a grizzly in the San Juans would again be confirmed by Ed Wiseman, an outfitter from Hooper, Colorado, when he was severely mauled by one in the southern San Juans near Platoro in 1979. Wiseman killed that grizzly to save his life, and it is now on display in the Denver Museum of Natural History.

Wildlife Conservation Officers

In the 1950s the demands upon the Game and Fish Department were changing and so the agency adapted to changes in the way it used its field people. Thomas Kimball became the director of the Game and Fish Department succeeding Cleland N. Feast in 1952. Kimball believed that the Game Warden position should be expanded beyond trapping and law enforcement duties. In 1954 he changed the position to a multipurpose job. The new title became "Wildlife Conservation Officer" (WCO). The change was rather traumatic to the Department's field force. A test was given and three-quarters of the men passed the test. A few quit, some retired, and some went back to college. In 1963 the job qualifications required a Bachelor of Science degree in wildlife management or some related field of study. The WCO now had responsibility for law enforcement, patrolling, investigating, fish management, game census, creel census, game bag checks, range surveys, live trapping and transplanting, game damage control and prevention, public contacts, speeches, sportsmen's meetings and more paperwork. There was little or no formal training except for teaching the new college boys how to trap and skin beaver and pack horses. A new officer learned about the real world from his experienced neighboring officers. Like many of the game wardens who came from the ranks of trappers and the game warden service in southwestern Colorado, Bill Schultz and Punk Cochran both passed the test and became WCOs in Creede and South Fork respectively. They were already old hands.

Some people consider law enforcement as the most important function of the Game and Fish Department, because it has the most visibility. In some districts the illegal killing of wildlife was a major factor for low wildlife populations; but law enforcement, although important, would not solve the problems of land abuse, population growth, and demand for wildlife recreation in Colorado.

The Upper Rio Grande didn't have a serious poaching problem when Bill Schultz was an officer.

Schultz said:

> Poaching was never serious around Creede. The worst poaching was downriver and around the San Luis Valley. A game warden was dependent on ranchers, hunters, fishermen, and other citizens who

wouldn't tolerate poaching. I think we have come a long way. At that time we had no contact with other officers like there is today. We had no two-way radio contact. You were strictly on your own. If you knew somebody was in a canyon and you suspicioned there might be someone poaching or bringing too many fish out, you could wait until they returned to check them.

A game warden had to earn trust by working hard, responding to people's needs and showing care for them as much as one did for the wildlife. It also helped to have a sense of humor. A warden was only as good as his grapevine. One needed many eyes to watch over big country. These new officers were soon involved in the many activities to perpetuate wildlife.

Schultz continued:

> We did a good job that few people were aware of. By the time World War II ended the antelope had been eliminated from the San Luis Valley. I was in Saguache at the time and I helped unload the first two truckloads over by the Mineral Hot Springs east of Villa Grove. We released the next load north of Saguache and then the next load was on the Curtis Ranch. Years later we put some north of Del Norte and some down around Romeo.

After the war the department was still using the fish stocking techniques of the 1930s. Old ways could not keep up with the demands of the increasing numbers of fishermen. New techniques came from men who were not satisfied with the status quo.

Stocking Fish

Schultz said:

> We made some big improvements in the way we stocked fish. We used to haul fingerlings in ten-gallon milk cans. We had to ice them down and we had to change the water every hour or so. We couldn't carry very many fish and we always lost some. Horses would spook at getting ice water sloshed on them and we had a few spontaneous rodeos. Then Norm Wilkerson, who was Superintendent of the Creede hatchery, developed the plastic jugs that worked so well that

Bill Schultz was a trapper, game warden, and Wildlife Conservation officer for nearly 20 years in the Creede District. He is shown here with his wife Eva and young Rob Hazzard (circa 1965). Photo from Lloyd Hazzard Collection

you could pack them horseback all day long and never lose a fish.

Norman Wilkerson solicited the help of Carroll Wetherill, local private fish culturist, and Carl Burnham (who owned Burnham Products Company, a plastics manufacturing firm in Wichita, Kansas). They fabricated a two-gallon polyethylene bottle that had a valve system that replaced airspace with pure oxygen. It was very simple process: a gallon of water, a pound of fingerling trout (800 to 1,000 fish), and oxygen added from a compressed oxygen cylinder. The bottle was then placed in an insulated bag to maintain a low temperature. Up to nine bottles could be carried on a pack horse. A couple of bottles could be easily backpacked to those "secret" beaver dams on public land. These fish bottles revolutionized the way fingerlings were stocked and were used by most federal, state, and private hatcheries for many years.

Charlie Kipp of the Wetherill Ranch unloading a bottle of fingerlings at Ruby Lake in 1968. Photo by author

In the early 1950s another fish distribution technique was being implemented as reported in the *Creede Candle:*

> With many chambers of commerce and sportsmen's organizations getting trout from the federal fish hatchery, Lake City has one of the most novel ways of stocking its mountain waters. The Hinsdale County Chamber of Commerce and the Lake Fork Sportsman's Association have been getting fish from the Creede federal hatchery and planting them from an airplane. Thousands of fingerling trout have been planted at Lake City. The last bombing of Devil's, Cataract, Crystal, Blue, and Slide Lakes in the Lake City district was for 36,000 brook trout. Rocky Warren and Bob Vickers of Lake City are pilot and bombardier for the project.
>
> On recent tests, a very high survival rate was shown on the trout dropped from planes. The time of getting them from the Creede hatchery and dumped into the lakes has been reduced to a matter of only minutes as compared to the old pack in method.

Game, Fish and Parks Department pilot Gordon Saville drops a load of trout into the Twin Ute Lakes early one morning in 1971. Photo from Division of Wildlife, Jim McFarland photographer

Lake City's Chamber of Commerce and other civic groups get the trout from the federal hatchery in the regulation manner of asking for them and guaranteeing stocking in specified waters. As a result of the stocking programs, fishing in many mountain lakes and streams of the Lake City district has been greatly improved over the past few seasons.

The Game and Fish Department started using airplanes to stock fish in the high lakes in the early 1950s. Norm Hughes was the pilot in southwestern Colorado who first flew a Supercub to a high lake, cut power, lowered flaps, and flew as low to the water as slowly as possible while a passenger poured a bucket of water and fingerling trout from the open door into the lake below. Experiments demonstrated that larger fish died on impact, but the fingerlings survived quite well. This stocking technique saved thousands of man-hours in hauling fish by horseback to the high lakes. All of the high lakes in the Upper Rio

Norm Hughes survived the crash of his Piper Supercub along the Continental Divide near the Twin Ute Lakes in 1961. Photo from Gene Bassett Collection

Grande were fairly easy to stock from an airplane, but some of the high lakes in the San Juans and Sange de Cristo mountains were so difficult and dangerous to stock by airplane that they would be stocked by helicopter in years to come.

Mountain flying can be very dangerous and there have been several Game and Fish Department airplane crashes. Gene Bassett, former wildlife officer in Bayfield, recalled:

> In November of 1961, department pilot Norm Hughes was up in the Ute Lake country and he got caught in the clouds and knowing his location radioed to the Durango Highway Patrol dispatcher that he was going to set the plane down at Twin Ute Lakes in the Upper Rio Grande. When he made his forced landing the wheels snagged in the deep snow and flipped the plane upside down just a few hundred yards east of Twin Ute Lakes. Norm was hurt only slightly. The Forest Service had a helicopter under contract from Skyhook Helicopters in Durango working west of Pagosa Springs. The dispatcher contacted

Rito Hondo Reservoir has been one of the most stable and productive fisheries in the Upper Rio Grande. Photo by author

that helicopter pilot, who had a Forest Ranger aboard that knew where Twin Ute Lakes were and they flew to the site and picked Norm up and flew him to Durango. Had they not been able to take him out, he may have died of exposure, even with the survival gear that he had on board. The following spring Errol Ryland, another Wildlife Conservation Officer in Durango, and I rode horseback to the Continental Divide and snowshoed over to the wreckage to pack out the radio equipment. The remains of that crash are still up there in a bog.

The Dingell-Johnson Bill

Funding for wildlife research and habitat restoration under the Pittman-Robertson bill of 1937 had been so successful that a companion bill was proposed for fisheries. In 1950 the United States Congress passed the Dingell-Johnson Bill that levied a ten percent tax on all fishing equipment. This bill provided funding for fisheries research, property acquisition, and the maintenance of purchased lands and waters. In the early 1950s dozens of small fishing lakes were built in Colorado. Trujillo Meadows Reservoir dam was built west of Antonito, Williams Creek Reservoir north of Pagosa Springs, Andrews Lake south of Silverton, and Denny Lake in Cortez.

Rito Hondo Lake was typical of the early Dingell-Johnson-funded fishery projects in Colorado. Located on Rito Hondo Creek, the site provided a lake of nearly forty surface acres. The lake never experiences winterkill, because most of it is too deep for significant aquatic vegetation growth and the creek flows all winter. Rito Hondo has consistently been a good fishery and is easily accessible. The problem with Rito Hondo was that the state built the dam, but had failed to obtain the water storage rights to fill it. The Game and Fish Department negotiated with Earl Brown to swap some water from the Upper Brown Lake to satisfy water users in the San Valley and the Rio Grande River Compact.

The Forest Service and Game and Fish Department built small dams on various streams around the state. Sometimes no more than a few passes with a bulldozer and a few loads of rock for a spillway made a permanent lake out of a beaver pond and created some popular fishing spots. The popular Spring Creek Pond located above the turnoff to Rio Grande Reservoir on Highway 149 was initially built as a small reservoir to supply water for the Wetherill Hatchery. Clayton Wetherill, grandson and namesake of pioneer Clayton Wetherill, remembered how that small pond became such a popular fishing spot: "In 1953 Mineral County Road Supervisor Bill Gustafson had my dad, Gilbert Wetherill, haul truckloads of dam building material to the site. I operated a D-6 bulldozer to increase the dam and size and depth of the lake." Mineral County has maintained the reservoir and the Game and Fish Department has stocked fish for the pleasure of many children and senior citizens ever since.

In addition to building new fishing lakes, the Game and Fish Commission directed the Game and Fish Department to purchase properties from willing sellers to increase the number of places where the public could fish. Road Canyon Reservoir had been a successful commercial operation for many years. The department purchased the Road Canyon Reservoir dam, a special use permit, and water storage rights in 1956 from the Bert Hosselkus estate. The lake had periodically experienced winterkill, but even though it had suckers and tench that competed with the trout, it had a history of having very good fishing.

The department purchased the Brown Lakes from Earl and Pearl Brown in 1958. These two lakes had also been successful commercial fisheries. At that time the lakes were very good fishing. They were only seven feet deep, but they didn't experience winterkill very often. However as the years went by the lakes began to fill with silt, and the frequency and severity of the winterkill increased.

In some cases the personal contact and diplomacy of field people paved the way for the state to purchase additional properties. Since the state could not condemn land and force a sale, it was up to the Wildlife Conservation Officer to contact landowners who wanted to sell their property to the state. Bill Schultz explained how the department bought the Coller State Wildlife Area:

> Punk Cochran and I heard from the Denver office that the department was buying property for elk range and fishing access so we went to Frank Coller who owned what is now the Coller State Wildlife Area on the Rio Grande above Elk Creek. He finally said that he wanted to sell it and that he would like the Game and Fish Department to have it so that it would never be closed to fishing. The property is an important wintering and migration route for elk along with three miles of the fishing access on the Rio Grande. Every time the Denver staff people had a meeting with Frank Coller they would call Punk and me and we all sat in. So after nearly a year Frank agreed to sell the property and it was all bought for fifty dollars an acre in June of 1959.

Jim Cochran described his father's ability to work with people for wildlife:

> Dad wasn't much of a dreamer. He just did things. He talked to people and ideas seemed to evolve. He could get cooperation from folks you wouldn't think you could get in the same room. Beaver Creek Reservoir was a major acquisition for the fishermen. When it was still an irrigation reservoir, Dad worked with the farmers to maintain the level as much as possible so fish would survive. I think dad's contacts in the San Luis Valley and the respect people had for him greased the skids for many of his projects. I never did really under-

> stand why farmers would give up that quantity of water for a few fish, but Dad was able to convince them to do so. The Game Fish and Parks Division purchased Beaver Creek Reservoir from the Mosca Irrigation District in 1965.

Just like other wildlife officers Punk continued to look for ways to improve opportunities for fishermen. Don Benson was Area Supervisor in the San Luis Valley when the Game and Fish Department started to develop Big Meadows Reservoir on the South Fork of the Rio Grande:

> Punk Cochran was instrumental in getting Big Meadows Reservoir built. He pursued his idea with Phil McCollough, State Water Commissioner for the San Luis Valley. The reservoir could not be built until the Rio Grande Compact could be satisfied. Phil's knowledge of water law at that time and Punk's ability to work with farmers in the San Valley was instrumental in clearing the way for the dam to be built.
>
> A post-compact reservoir could not be built, because the department had no water storage rights. After 1938 the Rio Grande Compact between Colorado, New Mexico, and Texas went into effect no more water storage rights could be granted in the Rio Grande Basin. Big Meadows Reservoir had to be filled with water from outside the Rio Grande basin. The department purchased the Tabor Transmountain Diversion on Spring Creek Pass. The Compact agreed to allow the Game and Fish Department to run Tabor Diversion water down the Rio Grande to satisfy the compact with New Mexico and Texas. The department was allowed to store the same amount of water in Big Meadows Reservoir. The legal transfer is made only on paper.

During this time wildlife officers, biologists, Forest Service personnel and many citizens in Colorado envisioned building small fishing lakes wherever there was a creek and a meadow. Most sites were not feasible, but once in a while one was found. It took much more work to see a project completed than the public ever realized. Punk Cochran got the idea to build the Big Meadows Reservoir near Wolf Creek Pass on one of his many horseback trips. He told the author:

> I was horseback coming out down from Archuleta Lake and sat there on my horse and looked across that valley and said to myself that if we could build a dam there it would make quite a nice lake and make some good fishing.

The department's engineering section completed the design in 1954, but the dam wasn't completed until 1967. Punk had to jump through the hoops of water law, engineering, and budgeting to convince people within the Game and Fish Department as well as water users and others. The Rio Grande Compact also had to give its approval. The Forest Service built a campground near Big Meadows Reservoir, and it became a destination vacation area for hundreds of visitors every year.

Punk Cochran was also instrumental in the department's purchase of Alberta Park Reservoir near the summit of Wolf Creek Pass from Edward Lohr in 1963. This small reservoir in a beautiful setting offered yet another opportunity for anglers in that area.

All of these new properties across Colorado increased the Game and Fish Department's workload and the state was slow to hire the workers to do the property development and maintenance. It was the Wildlife Conservation Officer who picked up the slack and painted outhouses, hauled trash, built fences, poured concrete, cleaned spillways, and enforced the use regulations. The purchase of Beaver Creek Reservoir near South Fork required Punk to make valve adjustments at Beaver Creek Reservoir to maintain water levels as directed by the state water commissioner. During the summer months his priority was to maintain the legal water level while his other responsibilities became secondary. Over the years the Division of Wildlife has invested hundreds of thousands of dollars repairing and maintaining these properties for people's enjoyment. Eventually, the Division of Wildlife was able to hire additional wildlife technicians to manage these properties.

Chapter 7

WILDLIFE, LAND, AND ATTITUDES IN TRANSITION

The 1960s brought another decade of change and growth to Colorado. People were discovering Colorado and moving here in greater numbers. The loss of habitat due to development was becoming serious. More information about wildlife was being demanded by legislators, developers, and concerned citizens. Major wildlife studies were being conducted by research biologists in the Game and Fish Department's Wildlife Research Center and the Colorado State University Cooperative Wildlife Unit in Fort Collins to increase our knowledge about the many aspects of wild animals, birds, and fish and their ecology. Later such studies would include amphibians and reptiles, but in this time period the focus was still on game species.

Society still placed high value on the utilitarian importance of wildlife, yet attitudes were beginning to change. Greater importance was given to the enjoyment of wildlife in addition to its food value. Managing lands and waters to maximize the numbers of fish and wildlife that could be taken by sportsmen remained a driving force behind most wildlife programs. People wanted more wildlife and it was the science that learned how to meet that public demand for the resource.

Fishing For Fun

Some traditions would change in response to new needs and ways of meeting them. Year-round fishing was one of those traditions that changed, beginning in 1962 when the Game and Fish Commission established year-round fishing. For many anglers, the loss of opening day of fishing season was like canceling Christmas. There were many reasons for not having an opening day, not the least of which was the fact that forty percent of the fish for the year were caught in a three-day period on opening weekend. Year-round fishing spread the fish out for more anglers over the entire year. Although year-round fishing had its critics, it gained popularity in western Colorado where one of the best cures for cabin fever was the opportunity to go out on a warm winter day and catch a few fish through the ice.

This same year the Game and Fish Commission established the first regulations that started a "fishing for fun" campaign by restricting fishing tackle, bag limits, and size limits on a few select waters. This restriction was not all that popular at first. The Coller Wildlife Area on the Rio Grande was one of those waters. Punk Cochran supported the regulation and took a lot of heat from some local merchants who saw the change as a threat to long established fishermen. Fishing for fun and fly fishing in particular hadn't caught on with many of the warm-water fishermen from Texas, Oklahoma, Kansas, and Arkansas. Even in the 1960s nonresident fishermen accounted for eighty percent of the fishermen on the Rio Grande and the Lake Fork of the Gunnison. Attitudes of both residents and nonresidents would have to change in the years to come if fish populations would be able to sustain the demand for recreational fishing.

Mixing State Parks and Wildlife

Harry Woodward came to Colorado in 1961 from South Dakota where he was director of the South Dakota Game, Fish and Parks. In 1963 the Colorado legislature created the Colorado Game, Fish and Parks Department. Since Navajo Lake State Park was the only state park in southwestern Colorado at the time, the name change didn't affect field operations in southwestern Colorado. Department personnel did become involved in licensing boats and snowmobiles, providing snowmobile operator training and enforcement of state park statutes. Park rangers checked fishing licenses and assisted Wildlife Conservation Officers. The new department had to work with multiple missions and keep parks and wildlife functions separate. The merger lasted until 1973 when complications arose over funding. State Parks were tax supported and the Game and Fish was funded by hunting and fishing license fees. If hunting and fishing license revenue was diverted for any other purpose than perpetuating wildlife, Colorado risked losing its Pittman-Robertson and Dingell-Johnson funding.

Elk Habitat Research

Richard "Dick" Denny and Ray Boyd, wildlife researchers, conducted the White River Elk Study in the early 1960s. This was the most extensive elk and habitat inventory study ever done in Colorado. According to Boyd:

> It appeared to us that there were some major differences between the White River and Rio Grande elk herds and we wanted to see if our thoughts were correct. We postulated that White River elk wintered on excellent oakbrush/grass/serviceberry winter range, while the Rio Grande elk wintered on almost pure grass winter range. The calf crop on the White River averaged about 65 calves per 100 cows and the Rio Grande calf crops averaged about 40 calves per 100 cows. Antler size on White River bulls appeared to be larger for each age class and the number of points on the antlers of two year old bulls and older bulls was about the same. Rio Grande yearling bulls nearly always were spikes and shorter than White River elk spikes. Yearling White River bulls often had additional points. These anatomical differences appeared to be attributable to nutrition, caused, we thought, because of poorer quality of the winter range forage in the Upper Rio Grande.

Denny and Boyd used new tranquilizer technology to capture about two elk per month in order to study elk anatomy and physiology. They measured various organs for comparison and found measurements and weights of vital organs of Rio Grande elk significantly smaller than White River elk. The pelage of Rio Grande elk was also noticeably lighter in color.

Boyd recalled how they learned a new banding technique:

> One day while tagging elk from a helicopter on the White River near Meeker, Dick Denney darted a cow elk. We landed near the cow and he finished ear tagging and neck banding it. All this time its calf would not leave so Dick just ran the calf down in the snow and ear tagged it.
>
> We both talked about this and wondered if we could just jump on an elk in the deep snow. We decided to try bull dogging elk from the chopper. Although some people criticized the technique, it was very effective. We didn't think mortality was much higher than with any other technique at that time and was perhaps easier on the animals in the long run. The main advantage was that immobilizing elk demanded that we stay with the animal until it recovered, so that we

Dick Denny, research biologist, stands on the skid of a chopper as pilot Virgil Jones maneuvers the chopper close enough to bulldog the calf elk below. The bulldogging technique was used only briefly as an experiment. Photo from Ray Boyd Collection

were able to do about three elk per hour. By bulldogging, we could do eleven elk per hour.

At first we just stepped off the skid onto the elk's back. I did this once and the chopper swung toward me after my weight was off the skid and my right leg got pinned under the cow and I tore some ligaments in my right knee. After this we jumped on the elk from about ten feet above them and about ten feet off to one side. This allowed the chopper to swing over our heads and no one got injured after we changed the procedure. We banded a total of seventy-two elk in the Upper Rio Grande using this technique in the winter of 1964-65. Most of the elk were tagged on Long Ridge and some in the river bottom.

Tagging these animals provided biologists with information about migrations. This was one of those behind-the-scenes projects that often goes unnoticed, but the project caught the attention of the national media. Boyd said:

Jim Fowler and Merlin Perkins came to Creede and filmed us bull dogging elk on the Coller Wildlife Area for one episode of Mutual of Omaha's popular television show, "Wild Kingdom." We tried to get Fowler lined up on one, but he never caught an elk. We had a ball working with him.

Boyd initiated an intensive helicopter elk census and the classification of bull and calf/cow ratios. Also a larger sample of elk was needed for determining animal weight and age information, and while the animals were captive they were ear tagged and neck banded to study migration patterns since biologists no longer bulldogged or tranquilized elk. Cliff Coghill, a Wildlife Conservation Officer in Gunnison, developed the elk trapping technique used throughout Colorado for many years. In southwestern Colorado elk were trapped in the Dolores, the Gunnison Basin, Hermosa, Bayfield, Pagosa Springs, and in the Upper Rio Grande. Bill Schultz and Punk Cochran started trapping elk on Goose Creek and Shaw Creek. Later elk were trapped on the Coller State Wildlife Area, Baughman Creek, Saguache Park and other sites around southwestern Colorado.

Elk were captured in a corral trap about 100 feet in diameter that was baited with alfalfa. The elk came into the trap at night for the free morsels and tripped a wire that closed spring-loaded doors. The door to the trap was designed to be narrow so that a large bull elk could not get into the trap. This usually worked, with a for few exceptions. The next morning wildlife officers would go to the trap and work the elk by first raising a window in one of the panels. Now theoretically an elk was supposed to jump to freedom through the "window" and the elk would find itself caught in a net that suspended it off the ground. The window was immediately closed to prevent another elk from making a dash to freedom. A blindfold was put over the head of the elk to shield the eyes and this quieted the elk. The net and frame were attached to scales and the elk was weighed, aged by dentition examination, ear tagged, neck banded, and released when the net was dropped to the ground. The information was recorded and the net reset for the next elk.

The theory was sound and literally hundreds of elk were banded with this technique, but there were a few times when the elk would not cooperate and would not jump. One such morning in the winter of

1966, there were about 20 elk in the trap up Goose Creek on the 4UR Ranch. Punk Cochran and the author surveyed the situation and Punk, being the veteran officer, went to his truck and got his lariat. The plan then changed to roping and throwing the elk to the ground, tagging, and aging, but foregoing the weight measurements. All the while the rest of the herd would patiently watch off to the side of the trap. Punk stood up on a ladder so he could reach over the eight foot wall panels. The elk began to run around the inside of the circular trap. Punk began twirling his lariat and announced that he would only catch the small elk just as his loop settled over the biggest and meanest cow elk in the San Juans. Quickly he dropped the rope and as the elk settled down and walked around the rope came loose and fell to the ground. Punk said: "I need my rope so we can try it again. Go in and get the rope. Take this broomstick with you and if an elk challenges you all you have to do is tap her on the nose." The author being young and taught to respect his elders took the broomstick and entered the trap. The rope was in the middle of the trap. As he got ready to reach down and get the rope the meanest cow elk of the San Juans jumped out of the herd to challenge. She started grinding her teeth to demonstrate her anger. His knees quivered. She started striking the ground with her front hoof. The coach who threw the rope in the first place yelled, "Grab the rope!" Just as his hand touched the rope the meanest, biggest cow elk of the San Juans got a tap on her nose, whereupon she instantly took a deep breath and barked like a huge, mean, man-eating dog. That sudden burst of blood-curdling sound precipitated instantaneous and indescribable relaxation of all the muscles in his scared-stiff body. Somehow through it all he grabbed the rope and crawled backwards to the exit. After a brief strategy meeting the door was wired open so the elk could leave in peace without their tags and bands.

Similar experiences caused inventive minds to modify new traps using a cattle squeeze shoot instead of a net to hold elk while they were being weighed and tagged. The newer traps had walls made of nylon netting and canvas curtains instead of plywood panels. Banding elk continues to have its exciting moments that are now reserved for trainees and volunteers.

Some neck bands lasted up to nine years, and one of the first elk ever tagged by Bill Shultz as a calf in 1965 was found dead in 1988 (23 years later) at Humphreys Lake by caretaker Bill Dooley. She still had

Punk Cochran puts a neck band on a Goose Creek elk. Sightings of these elk were reported and biologists determined the extent of migrations. This information was necessary to establish logical game management units so that elk populations could be better managed. Photo by author

her ear tag. Her teeth were worn to below the gum line, but she had calved that spring. After all those years she died only two miles from where she had been banded. The Rio Grande Elk Study showed biologists that elk which wintered on the San Juan side of the Continental Divide moved to summer ranges on the Rio Grande. Only a few elk migrated northward toward the Gunnison basin. Most elk that wintered between South Fork and Goose Creek spent the summer along the Continental Divide as far west as Trout Creek. This may have been started as a result of elk being released on Elk Creek and Goose Creek in winter months and moving to high country in the summer to escape the biting insects and feed on the lush sedge marshes of the high country. Elk that spent the summer in Lost Trail Creek, Finger Mesa, and North Clear Creek tended to migrate to winter grounds down toward Long Ridge below Bristol Head Mountain.

Elk migration information was gathered throughout western Colorado and used to refine management unit boundaries that were based on biology rather than political divisions that animals don't respect anyway. Banding information verified that in the mid-1960s greater numbers of elk were migrating from the Coller State Wildlife Area to drainages north of Del Norte. Punk Cochran considered Old Woman's Creek the safety valve for Rio Grande wintering elk, because elk seldom used to be there except in the most severe winters. Soon elk were increasing in that unit and stayed there year around. All this information would become important in later years when the Division of Wildlife established a quality elk hunting unit in the Upper Rio Grande.

Grizzly Bears and Mountain Lions

The legislature officially protected grizzly bear in 1964 more than a decade after the "last" grizzly had been killed in the San Juans. The Grizzly Bear Management Area south of Rio Grande Reservoir was dropped from the hunting regulations. There were unconfirmed grizzly sightings from sheep herders, hunters, and other outdoors people almost every year. Another grizzly search was started in 1970. The Game, Fish and Parks Department bought some old horses that were going to be slaughtered. Wildlife officers Gene Bassett and Herb Browning, from Bayfield and Pagosa Springs respectively, along with the author led these horses to potential grizzly habitats along the Continental Divide where they were killed as bait and were monitored. They operated on the theory that only a grizzly was large enough to move a horse carcass.

Herb Browning, said:

> Besides using the airplane to monitor the carcasses we used an automatic camera that would take a picture anytime something walked into view. The first time I checked the camera I was excited that all the film had been used. When I had the film developed I was disappointed to see only pictures of magpies jumping on and off the carcass. One camera did, however, take pictures of thirteen different black bears on one carcass. We didn't find any grizzlies, but we learned that we had more black bears than previously thought.

Other search techniques have been used and each one has its merits, but the San Juan grizzlies must have become very adept at avoiding humans. (To say there were no grizzlies remaining would have been inaccurate, because at least one grizzly was still alive in the South San Juan Wilderness near Platoro, Colorado.)

The legislature removed the mountain lion bounty in 1965 and declared it a big game animal. Removal of the bounty had no impact on lions in the Upper Rio Grande, since there have never been many mountain lions in the area. Although there are excellent mountain lion habitats in southwestern Colorado, the bounty systems have always been ineffective means of controlling predators, because they do not necessarily remove damage causing animals. Such a system is usually corrupted by individuals from other states who kill lions and bring them to Colorado to claim the bounty payment.

The Human Impacts

In the early 1960s Colorado's human population was exploding and wildlife habitat was shrinking at an unprecedented rate. Many new residents were drawn to southwestern Colorado's outstanding hunting, fishing, scenery and climate. This was also the last mountainous area of Colorado for developers to discover. The region had always been too distant from population centers, but modern transportation brought more people who fell in love with the San Juans. Dramatic population growth and subdivisions and related development began to fragment historic and critical wildlife habitat that threatens the future survival of wildlife.

Silverton's and Del Norte's populations grew in the 1960s as silver and gold mining exploration and mining brought in more people. The silver mines near Creede were active again in the 1960s and the permanent population grew from about 350 to nearly 1,100 by the 1970s. In the early 1960s there was only one subdivision south of the Creede airport, and there were no year-round residents; but soon other subdivisions would be platted up and down the river in prime wildlife habitat.

Subdivisions from South Fork to above Creede gradually forced deer and elk to find less desirable places to winter. In the severe winters of 1964, 1968, 1978, and 1984 elk were forced to the edges of subdivisions and even to the Creede city limits trying to find forage. In 1964 there were elk that came into Durango. Any drainage that had a southern exposure, some escape cover such as timber nearby, and any vegetation, was critical elk and deer winter range in the Upper Rio Grande. Unfortunately for elk and deer these areas are also prime homesites. This devastation to wildlife is permanent and cannot be reversed by closing seasons or making refuges. This is further complicated by the grazing of cattle in the summer on shrinking elk and deer winter range. Following a dry summer there is little natural feed left on the National Forest winter range so elk and deer are forced onto private lands to meet their nutritional needs. This shrinking of habitat has been taking place around Lake City, Pagosa Springs, Bayfield, Durango, Mancos, Dolores, Ouray, and Gunnison as well as many other areas of Colorado.

The Colorado Legislature passed House Bill 1041 in 1974 to empower counties to regulate development. Each county appointed a land use administrator to assist county commissioners to consider environmental issues in their planning process. Charles Steele was Mineral County's first land use administrator. Mineral County has only about five percent private land, but those lands are often critical wildlife habitats. Steele worked with developers and agencies such as the Soil Conservation Service, Game, Fish and Parks, and others to develop subdivisions that have been more considerate of wildlife values than in many counties in Colorado. The Phipps family were the first landowners in the Upper Rio Grande to donate development rights by way of a conservation easement to The Nature Conservancy so that the La Garita Ranch will never be subdivided. Such early foresight set an example for landowners to capitalize on the value of their land without developing it. It assures the land will remain open to protect the natural environment, its scenic character, and valuable wildlife habitat. Throughout southwestern Colorado counties and private organizations are working together to purchase land or gain conservation easements to protect critical wildlife habitat before development destroys it forever.

Wilderness

Congress passed the Wilderness Act in 1964. As a result of this landmark legislation the Weminuche, South San Juan, La Garita, West Elk, Big Blue, Sneffels, Lizard head, and Powderhorn Wilderness areas were established in southwestern Colorado. To implement this new legislation the Forest Service began a process to evaluate roadless areas that could be considered for wilderness designation. In the case of the Weminuche Wilderness, the original Upper Rio Grande Primitive Area on the Rio Grande and the San Juan Wilderness area were combined along with adjacent lands to create the largest wilderness area in Colorado.

The process of gathering public input was wide open and all sorts of recommendations came in. The timber industry proposed a wilderness boundary that would extend from the alpine tundra down to timberline. All areas that included any marketable timber or potential for such would have been excluded in their proposal. The Forest Service's fifty-year road plan map proposed logging roads on most ridges and valleys so that all timber could be harvested. Had such a plan been implemented there would be no escape timber for elk or deer and no mature forest to sustain a diverse ecosystem. If the aftermath of logging activity in the Del Norte Ranger District in the early 1960s was any indication of what was in store for the rest of the Upper Rio Grande, it was apparent that the future for elk was bleak.

In 1968 Game, Fish and Parks Department personnel were assigned to develop a map and narrative that became the department's recommendation for a wilderness boundary for the Weminuche Wilderness. The department's position was based strictly on wildlife impacts and law enforcement considerations. There wasn't time to do scientific studies on which to base recommendations, but they had to be made in the knowledge of what was happening to wildlife in the Del Norte District and some of the massive clearcuts already completed on the Creede District. One didn't need to be a rocket scientist to see the damage that logging and road building was doing to wildlife habitat. The department sponsored "show me" trips and pack trips for legislators and others to let them see first hand what would happen to the back country of the San Juans if the management programs proposed at that time were implemented in this pristine environment.

Mike Zgainer, who was the Wildlife Conservation Officer in South Fork District said:

> The Forest Service's clearcutting of the South Fork and Del Norte peak country was totally the worst thing that ever happened to wildlife, soils, water quality, and it will take 100 years for much of that

country to recover. A lot of the timber stands were "high graded" by cutting most of the timber, but only taking what they wanted."

The Game, Fish, and Parks Department was concerned about the proliferation of roads into the back country. The Forest Service policy was emphatic that once a road was constructed it had to remain open, and there was no way the Forest Service could close roads that had been paid for at public expense. So the alternatives were limited-wide open or total wilderness. If there had been some flexibility and creativity, the Weminuche Wilderness might not have been so large. A congressional hearing was held in the Creede gymnasium so that the San Luis Valley community could speak to the representatives. Don Smith, Wildlife Program Specialist for the department, testified about the impacts that wilderness designation would have on wildlife. Smith had compiled all Wildlife Conservation Officers' recommendations and his own notes into a Game, Fish and Parks Commission position statement that was incorporated into the Environmental Impact Statement. The Mineral County Commissioners were the only such local commission that supported the creation of the Weminuche Wilderness. The final boundary was a compromise.

When the Weminuche was mapped and publicized, the new wilderness was invaded by backpackers and outfitted groups who came to see the largest and most spectacular mountain wilderness in Colorado. Wilderness itself is not the panacea for protection of wildness. Legal designation in and of itself doesn't protect the wildness of land. Just limiting modes of travel to walking and horseback riding doesn't maintain solitude and the wilderness experience. Uninformed wilderness users can have devastating impacts on some high mountain lakes, sensitive alpine and riparian areas, and the critical nursery areas of elk. Nevertheless, the Weminuche Wilderness protects some of the wildest and most isolated land left in the state of Colorado.

The La Garita Wilderness north of Creede is on the Gunnison and Saguache watersheds. When it came up for designation, some wilderness advocates recommended that the wilderness boundary extend southward to include Wheeler Monument, Wason Park, and down to Colorado Highway 149, private land and even the Creede dump. The timber industry based its proposal on timber harvest con-

siderations. The Game, Fish and Parks Department put its recommendation together for the La Garita Wilderness in the same way that it had for the Weminuche Wilderness proposal. Bob Hoover, the Division's Wildlife Program Specialist, took a pack trip and spent many days on foot, horseback, and in a helicopter with the author investigating alternatives for wilderness. Hoover compiled all the field information that eventually became the Game, Fish, and Parks Commission recommendation to the Forest Service. As a result, that portion on the Rio Grande side of the proposed wilderness was not included in the wilderness, but was to remain roadless, except for a vehicle access corridor to the Wheeler Monument.

The author surveys the beautiful Ute Lake Basin in the heart of the Weminuche Wilderness Area. (circa 1983). Photo by author

Every wilderness went through the same process. Each was unique, but had common issues. The same concerns and issues were faced throughout Colorado as conflicting values and interests fought over the future of western Colorado's mountains, forests, and water. In most cases there was compromise, and few got all the concessions they wanted.

Wildlife Conservation Officers Gordon East and Dan Potts seine cutthroats for spawning at Haypress Lake. The fish hatched from this operation were stocked throughout the San Juan and Sangre de Cristo's high lakes and streams. (circa 1968). Photo by author

Seeking More Ways to Increase Fishing

In the 1960s the Colorado Game, Fish and Parks Department tried to stock more cutthroat trout into the high country, but a reliable source of eggs was not always available. In 1963 wildlife officers Punk Cochran, Bill Shultz, and supervisor Don Benson negotiated an agreement with A.E. Humphreys to take native cutthroat spawn at Haypress Lake. The Humphreys family has maintained a cooperative agreement that has produced millions of cutthroat eggs ever since. The department experimented with several subspecies of cutthroat at Haypress Lake, including the Yellowstone, Snake River, and Colorado River subspecies. In the 1990s the Division of Wildlife established the Rio Grande cutthroat brood population at Haypress Lake which met with limited success until a kidney disease infected the population and spawning operations were halted.

John Alves, fish biologist, gently spawns a female Rio Grande cutthroat. (circa 1995) Photo by author

The process of spawning wild fish was simple and not unlike that used by the pioneer fish culturists. Every spring department personnel dug pools for fish traps on a tiny inlet stream. The cutthroat were trapped and put into live boxes in the lake. A team of wildlife officers, biologists, and fish culturists stripped the eggs from females and fertilized the eggs with the milt of the males. The eggs were "hardened" for an hour or more and then transported to the Creede hatchery.

When the eggs hatched and the fry had grown into small fingerlings (usually in early September) they were transported by pickup truck, horseback, backpack, airplane, and helicopter to be stocked in back country lakes, streams, and beaver dams throughout the San Juan and Sangre de Cristo ranges. The department started using heavy-duty plastic bags in place of the fish bottles that Norman Wilkerson had developed. The plastic bags were filled with a gallon of water and a pound of fingerlings. The bag was inflated with compressed oxygen and sealed with rubber rings. The plastic bag was then placed in heavy cardboard boxes or fish food sacks and, like the

The author and Bill McWilliams of Big River Guest Ranch watch fish culturist Ray Ayr unload fingerling rainbows in the stream-side pond at Cottonwood Cove. These fish were fed all summer by ranch staff and released into the Rio Grande in the fall. Photo by author

plastic fish bottles, they could keep fish alive all day. It took thousands of man-hours every year for local outfitters, volunteers, and wildlife officers to pack fish into the back country. This practice continues today.

The stream-side pond program had been in existence since 1918 but gained momentum at the beginning of the Depression, because fish food and transportation were expensive. It was thought at the time that it was better to allow tiny fry to grow larger on natural feed before they were released to a river. This program was still functioning into the 1970s. Over the years the department had stocked thousands of fingerlings into these ponds. In the fall the landowners released fish that had grown up to seven inches into the Rio Grande. Sometimes keeping a tradition makes people feel good, but the effort may not be accomplishing its purpose. A fall release of domesticated pond fish left little time for them to acclimate to a wild flowing habitat with natural food. Few of these fish found a habitat where they could survive the winter. It was also true that wild trout such as the brown are well established in most Colorado rivers and had an affinity for gobbling up the small fish that entered their domain.

In the late summer of 1970 Darryl Todd, a graduate student working for the Phipps Ranch, Ranch Manager Dan Anderson, and the author tagged 1,000 small rainbows which were raised in the La Garita stream side pond in order to evaluate the effectiveness of the streamside pond program. A total of 10,000 fish were released and only one tagged fish was ever caught and reported. It was obvious that for all the investment of fish, feed, and time, the fish were not getting to the angler's creel. The streamside pond program was suspended.

While working on the La Garita Ranch, Todd was also working on his Masters Degree through Colorado State University. Todd evaluated the impact of stream improvement structures that the Phipps Ranch had installed in Bellows Creek. In the process he discovered a species of algae that lives only in high velocity streams that have temperatures just above freezing. Such information added knowledge about mountain streams

One other attempt was made in 1930 to improve the population of wild fish by Game and Fish Commissioner R. G. Parvin when he closed some small streams around the state such as Farmers Creek and Blue Creek on the Upper Rio Grande. These were considered prime streams for wild trout to migrate to and successfully spawn. It was hoped that the small trout that hatched would swim downstream with the next spring runoff and increase the trout population in the river. That program was soon suspended.

Winter Feeding—The Two-Edged Sword

Winter is the most difficult season of the year for both man and animal in Colorado. Elk survive amazingly well during most winters. (Deer, as mentioned before, do not fare as well in the marginal habitat of the Upper Rio Grande.) Once every few years however, severe winter conditions can become deadly for elk-1964 was one of those devastating winters that would bring change to the Game and Fish Department's policy toward winter feeding of big game. Ed Hargraves, former postmaster of Creede, said:

> The year all this started we had a big wet snow, then some really cold weather and then more snow. There was a layer of ice that the animals couldn't paw through, and after seeing some animals die, the citizens of Creede started to feed. Gene McClure and I bought the hay at first, but after the newspapers got in the act, we had more hay

donated by San Luis Valley farmers than we knew what to do with. We had people unloading trucks at night, people with snowmobiles and sleds that fed animals away from the roads. We hauled salt blocks and a pellet used for cattle. Dr. Leary a veterinarian in the Valley asked us to stop using them because they were too rich. The county plowed out places where the animals were concentrated and a person would take that area as his responsibility so there was no duplication of feeding. We really had more help than we could use.

When we started to feed, Bill Schultz asked us to stop because it 'was not Department policy.' I called Harry Woodward, the Director, and he also told me to stop. We asked for a meeting in Creede, but he had one in Monte Vista. A delegation from Creede went down, but it didn't last long. They accused us of harassing Woodward.

The Game and Fish Department took an unpopular position and paid a public relations price for putting biology ahead of public sentiment, as reported in one of the local newspapers:

Mineral County War Over Elk

by Pasquale Marrazino

The Creedites came through as they have for almost 100 winters. Ed Hargraves, who is postmaster and runs the Creede Hotel, and Gene McClure who runs a general store, bought up tons of hay about 10 days ago and began jeeping it to the herds.

Many elk have died. The condition of the herds was poor. Many of the animals and especially the calves were dropping from starvation.

. . . Harry Woodward, Department Director, sent along a demand for unconditional surrender, ordering the Creedites to stop all feeding of the animals.

Their answer was a little more ungentlemanly than that given the Nazis by Gen. McAuliffe at Bastongne. They told Woodward and his entire Department to go to hell.

The stand brought support from ranchers in the San Luis Valley. They dug into their hay stores and trucked them into the hungry herds around Creede. That brought new demands to stop feeding from Game and Fish.

"They can make as many demands as they want to," Hargraves said. "We looked up the Colorado statutes on game preservation and the law specifically says that wild game must be fed by the department under orders from the Governor. We haven't appealed to the Governor, but he can see the situation for himself. If he cares to act, fine. If he doesn't we'll care for our own game."

Not all Department personnel agreed with official policy, especially the field personnel who had to take the heat on a daily basis for the department not giving in to public demand.

Johnny Jackson remembered that meeting in Monte Vista "'Smoky' Till, the Regional Manager, walked up to me and slipped me a twenty dollar bill and said, 'Here, add this to the kitty.' I thought Smoky was one of the best."

Harry Woodward explained:

> When I came to Colorado winter feeding was not a hot issue until 1964. The staff at that time had overseen a number of failed feeding operations. I had also been involved in South Dakota with failed winter feeding operations. The staff recommended against feeding and the commission backed the biological and technical position of staff. It was my job to follow the instructions of the commission. Like any director I caught the flak for supporting such an unpopular position.

The department's "let them starve to death" attitude was a source of contention and distrust for many years. Many elk were undoubtedly saved, yet some were killed by the feed itself, which they couldn't digest. The citizens did the job and to their credit saved many elk from starvation, but no one really objectively evaluated the effort. There have been many winter feeding operations in Colorado, some on an even larger scale, but most had failed and were very expensive.

There are two rules of winter. Number one is that some animals die of starvation and disease. Rule number two is that humans aren't going to change rule number one. Feeding animals doesn't cure them of pneumonia, battle wounds, disease, or old age. Feeding elk and deer perpetuates animals that, for the well being of the herd, should be allowed to die naturally. For instance, saving the late-born female calves produces more cows that tend to have more late calves, which do not survive harsh winters. Leaving nature alone in this respect would be better (although seemingly cruel and heartless) for the pop-

Volunteers feed elk in Sawmill Gulch west of Creede in February of 1968. Photo by author

ulation, but the public doesn't accept this two-edged sword.

There are the very mild winters when nearly all animals survive. When the first snowfall of the next winter arrives, some animals that would have died in an average winter are going to look skinny and some well meaning person demands that the department start feeding the whole herd. If the department is slow to start feeding, then the public gets the perception that the agency is irresponsible in its mission to protect wildlife. No one wants to see animals suffer so they feel a responsibility to help. If the animals are going to die, the public would rather see them die with a full belly, no matter the cost.

In 1968 another severe winter came to the Upper Rio Grande. About 150 head of elk were isolated between Shallow Creek and Miners Creek and were not able to move in the deep snow to find forage. Since the 1964-65 winter the Department Research Center in Fort Collins had been working on a food supplement that elk would eat and survive on. Bob Keiss, Wildlife Researcher, shipped ten tons of pellets which he had made in Fort Collins to Creede. Smoky Till, Don Benson, and the author met with Creede Rotarians and told them up front that there was a localized problem in Miners Creek and that the department was going to do something about it. No demand was ever made to start feeding elk throughout the district.

The department bought a Kristi Snowcat and started feeding elk. The only problem was the author had no experience feeding elk pellets. The first trip he spread out a couple bags of pellets on a packed trail and the elk wouldn't eat them. Charlie Kipp of the Wetherill

Ranch suggested putting some pellets on top of scattered alfalfa and the elk would learn it was food. The next day the elk consumed all the pellets that were mixed with the alfalfa. Soon volunteers rode in the sled and scattered sheaves of alfalfa and poured pellets on the packed trail every day. Toward the end of the winter a few mature elk died on the feed grounds. The pellet was too "'hot." Some elk got hooked on eating the pellets exclusively and died. Elk that ate a mixture of alfalfa and pellets survived. The late calves all died. Overall, the elk population survived throughout the Upper Rio Grande. Conditions in the rest of southwestern Colorado were not as severe and the department didn't do any winter feeding. As a result of this feeding experiment, the department continued to research developing of artificial rations for deer and elk that could be digested. More harsh winters were ahead and the public would become even more emphatic about winter feeding.

Chapter 8

WATER FOR WILDLIFE

The San Juan Mountains act like a gigantic snow fence and catch the snow laden clouds of western storms all winter, holding the snowpack until spring. The spring runoff is gathered by tributaries of the Colorado River from the West Slope of Colorado that flows down to the Pacific Ocean. Beginning at Stony Pass the mighty Rio Grande gathers its tributaries and flows down to the Gulf of Mexico. With all this water one would hardly think of southwestern Colorado being the arid place that it is. As in all the major watersheds in Colorado, the demand for water for domestic, industrial, agricultural, and hydroelectric use is in most years greater than the supply. The Colorado and Rio Grande Rivers have compact agreements with downriver states that require these river basins to deliver minimum volumes of water to those states. Agriculturists built dams and diverted water to irrigate crops. To improve rangelands ranchers built small reservoirs and developed existing springs in arid sections of western Colorado to water their livestock. In some instances their efforts also increased the number of deer or elk in those areas. Natural ecosystems in these areas were changed by putting water where there had been little or none, and thereby increasing the grazing of forage that grew sparsely on lands that were susceptible to soil erosion. Agriculture in some cases created fisheries in the form of reservoirs, but often at the cost of degrading river systems downstream.

In order to perpetuate fish and wildlife, the Game, Fish and Parks Department entered the water management business to intensively manage some remaining critical terrestrial and aquatic environments. Wildlife need water, not only to drink and to sustain vegetation for food and cover, but of course for the aquatic species water is home. Fish need good water quality and trout especially require higher than average dissolved oxygen levels to sustain them.

Preventing Winterkill

In southwestern Colorado most of the private lakes and some of the natural lakes in the high country have man-made dams. These reservoirs were built to store irrigation water for ranchers, farmers, and for commercial fisheries. Some of these reservoirs were shallow and soon after construction they began to silt in. The density of aquatic vegetation and algae soon choked these lakes top to bottom. When the winter ice cap formed, deep snows cut off sunlight to all this biomass and the bacterial process of decomposition depleted the dissolved oxygen. The fish grew rapidly in these productive waters in the summer but in the winter died from asphyxiation, unless there was sufficient fresh water flowing into the reservoir. For many years people have tried to reverse this natural process. Bert Hosselkus developed Road Canyon Reservoir into a fishery in 1908 and after a while it started to winterkill as the April 19, 1930, *Creede Candle* reported:

> **Several Lakes May Have Frozen Out**
> It is reported that Road Canyon and several others of the fish lakes in the Rio Grande Forest areas show a good many dead fish around the edges of the lakes where the ice has begun to disappear. At one place 47 dead and one live trout were counted. It is believed likely that several of the lakes may have frozen out thru the winter. Such freeze-outs are caused by the freezing up of the small creeks that feed the lakes. The creeks freeze solid and no water runs into the lakes and they also freeze solid to the bottom, killing the fish.

Contrary to what was written in this article, most lakes do not freeze solid. Seldom does the ice get more than three or four feet thick. Bert Hosselkus felt that if Road Canyon Reservoir fish survived every other winter he could make a financial success out its commercial fishery potential. Hosselkus tried to prevent winterkill at Road Canyon Reservoir by aerating the lake. In 1932 Hosselkus set up a big tractor on the dam. He stayed there all winter and ran a tractor-powered water pump. He pumped the water onto the dam and splashed it against a big board. Then the aerated water flowed back into the lake. The effort kept some fish alive, but the rest of the lake winterkilled. Bert built the upper Road Canyon Lake, that is now a waterfowl viewing area, to grow fingerlings. Each fall he opened the valve and drained the lake to release the fish that had grown through the summer and they were carried the short distance to Road Canyon Reservoir. He also thought he could prevent winterkill by releasing water from the upper lake in an effort to "freshen up" the lower lake. That didn't prevent winterkill either.

Hosselkus tried opening the outlet valve at Road Canyon Reservoir to suddenly drop the ice cap and crack the ice hoping to get air to the water and release the gases such as methane and hydrogen sulfide. The only problem was that by the time one could smell the hydrogen sulfide the lake had probably already winterkilled.

Wright's Ranch lower lake at the headquarters was excellent fishing, but soon after construction, began to winterkill, according to Howard Kennell:

> Carroll Wetherill and I took a million brook eggs at Wright's lower lake. We built the middle dam with V handled shovels and a team of white horses. Wallace Wright built a check dam there so we could trap brook that were about two pounds and then that lake got to winterkilling. Wallace got the idea that if he built an upper dam he could divert water down to the lower lake and save those big spawners and it wasn't long until they lost their water right. They lost all the fish in that lake and they gave it up.

Winterkill became a problem in several shallow lakes. People were willing to try almost anything to combat aquatic vegetation. In 1928 the Creede Chapter of the Izaak Walton League passed a resolution to have Seepage Creek lakes stocked with turtles to clear out the moss. There is no record of turtles being stocked in Seepage Creek, but chances are they would not have survived a San Juan winter.

Leroy Brown saw winterkill becoming an increasing problem at the Brown Lakes:

> The Brown Lakes partially winterkilled. I never remember a total freeze- out where a lake froze solid. The only thing that kept the fish alive was the South Clear Creek from that upper country. When the silt started washing down it spread out at the upper end of the lake and when the winter freeze would come, the creek bottom froze and the water spread out all over that upper end and would go out on top of the ice. That is when it started to winterkill.

Electric powered Fresh Flow Aerators aerated water and kept small areas along the shoreline from freezing, but were unable to oxygenate enough water to sustain trout. (January, 1968). Photo by author

In 1951 Charles Sickles devised an aeration system at Hermit Lakes. He mounted a thirty-three-horsepower outboard motor to a venturi air jet, but as with most attempts, the volume of water and the duration of the operation were not enough to satisfy the lake's demand for dissolved oxygen. The lake winterkilled.

Since decomposing aquatic vegetation is the major contributor to oxygen demand, it was worth an attempt to chemically remove the moss. In June of 1961 the Game and Fish Department spread sodium arsenite on the waters of Road Canyon Reservoir to kill the aquatic weeds. However, a short time later cattle were turned onto the nearby range and grazed on the grass in the shallow water, killing seventeen head by arsenic poisoning. As a result the department paid for some high-priced cattle and didn't solve the winterkill problem either. The chemical treatment did little to eradicate the aquatic vegetation and winterkill continued to be a problem.

A few years later the Game, Fish and Parks Department spread coal dust from an airplane on Road Canyon to attempt to increase the absorption of heat from the sun and melt the snow for the purpose of increasing sunlight penetration of the ice to increase photosynthesis. Soon after the coal dust was applied the next snow storm covered the coal dust and the idea was not tried again.

Lloyd Hazzard, Regional Fish Biologist, initiated a research project to prevent winterkill in 1966 when an electric line was constructed past Road Canyon Reservoir. Hazzard tested three stations with two Fresh-Flow aerators each. Used in sewage lagoons, these pumped lake water from the lake bottom, aerated it and sprayed it back into the lake. However, the fish which were attracted to the open water in front of the aerator died, because the pumps could not aerate the water enough to sustain them due to the high oxygen demand in the lake.

In 1970 Bill Babcock, a Game, Fish, and Parks Fish researcher, continued experiments and research on winterkill. Babcock really wanted to improve the fishing, and he was a risk taker always looking for ways to save winterkill lakes. The first aerator that Babcock experimented with was called the "Air Aqua System." This system used six one-half horsepower air compressors that pumped air into a lead-keeled plastic pipe with needle point holes that produced very tiny bubbles into the water. The bubbles rose to the surface, oxygenating and moving warmer bottom water which melted large sections of ice cap. This system melted three sections of ice about three hundred yards long. The winter winds came down the canyon and whipped the surface into white caps and aerated the water. However, this system was constantly plagued with mechanical problems that required almost daily maintenance.

Babcock's second and most successful system was a Helixor aerator. The Helixor was a fiberglass tube nine feet tall, eighteen inches in diameter with a spiral baffle corkscrewing lengthwise. Air supplied by a compressor was pumped to the bottom of the vertical tube and air rose to the surface pulling large volumes of water off the bottom of the lake, aerating it, and the warm water melted the ice cap.

Because the Helixor aeration system proved superior, when Babcock concluded his research, the Division of Wildlife permanently installed a small compressor building near the dam and buried pipelines to the underwater Helixors. This system has successfully prevented winterkill most years, but there are years when even this system cannot meet the dissolved oxygen demand of the lake and fish will winterkill.

The Helixor aeration system oxygenates large volumes of water and melts the ice cap at Road Canyon Reservoir. It has successfully prevented winterkill for more than two decades. (circa 1972). Photo by author

Like many winterkill lakes in Colorado, Regan Lake at the headwaters of House Canyon had no electric line nearby. It is a fluctuating reservoir, has a very small watershed, and no water flows into the lake in the winter. The lake is normally very shallow and produces fresh water shrimp and fathead minnows in tremendous numbers. As with all winterkill lakes, newly stocked fingerlings grow rapidly, but die in their first winter. If this lake could be saved, it would make an excellent fishery.

The Game and Fish Department's first attempt to prevent winterkill at Regan Lake was in 1956. They constructed a diversion structure to catch spring and seep water above the lake and put it into a collection pipe that flowed into the lake to provide oxygenated water. The volume was insignificant and proved ineffective.

Since there are a number of isolated winterkill lakes in Colorado that have no electric power nearby, Babcock designed and built two windmills that powered air compressors. However, equipment that worked well in summer tests did not work as well in sub-zero temperatures. High winds and severe cold temperatures broke the cast iron parts of the windmills. On several occasions Ralph Conkey, caretaker at Rio Grande Reservoir and an expert welder, volunteered to snowmobile into Regan and braze the pieces back together. Even though small areas of open water could be maintained during windy periods, they would freeze over as soon as the wind stopped. Although the technique could have application in other locations, the project was unsuccessful at Regan Lake and was terminated.

When the wind-powered systems proved ineffective in preventing winterkill, the department experimented next with a gasoline powered air compressor. Such an operation required an attendant, and the old Regan homestead cabin served a useful purpose in keeping volunteers warm and dry. In spite of the effort, the experiment failed. Babcock concluded his aeration studies in the early 1980s. Some said that Babcock's bubble machines made him the Lawrence Welk of the fish world.

Bill Babcock and two volunteers service one of the windmills at Regan Lake. This location did not have sufficient wind frequently enough and as a result the windmills could not prevent winterkill. (February, 1973). Photo by author

Lake City resident Vernon "Tubby" Carl built a wind powered air compressor at his own expense which he installed at Devil's Lake north of Slumgullion Pass. As a private citizen Carl was constantly investigating and inventing new techniques to try to save the fish in a beautiful alpine lake. He had good success and after his death his son Tommy continued to service the equipment and keep the lake alive. Now the Bureau of Land Management is forcing Carl to dismantle the aerator, because Devil's Lake is now in the Powderhorn Wilderness Area where man-made structures are forbidden.

One of the best techniques to prevent winterkill is to divert a stream into a lake. Brown Lakes had been very productive fisheries, but after the state bought them the trout winterkilled frequently and almost completely. Jim Landon, Regional Engineer, designed a pipeline that diverted water from Porcupine Creek (which normally flowed into the lower Brown Lake) and diverted it to the upper lake. Adding additional water flow to the upper lake increased the volume flowing over the outlet spillway into the lower lake. As it turned out, the fish in the upper lake that moved to the area of aerated water lived, or were caught by fishermen, but fish outside that small area of influence died.

Since the Brown Lakes started to winterkill when silt filled the South Clear Creek inlet, the author decided to try another approach. The author bought a case of ditching dynamite at the local hardware store and hauled it up to Brown Lakes. One fuse and one blasting cap were put in the first stick. Ditching dynamite is so pressure sensitive that when that first stick exploded it detonated all of the dynamite sticks that were pushed into the marsh eighteen inches apart. When that single blasting cap exploded a channel eight feet wide and six feet deep was created in a thundering instant. The experiment worked, but it became apparent that to clear a quarter of a mile of channel through a bog was nearly impossible. It also became apparent that the blasted channel would soon fill in with silt and require periodic removal. The cost of such a project would be prohibitive.

Unwilling to give up, the Division of Wildlife took what it had learned from Babcock's research and installed aeration systems in the Brown Lakes. Division personnel installed aerators similar to those that were successful at Road Canyon. The system was modified several times, but after trying and failing for eight years the aeration system was removed. Many shallow lakes that have been popular fisheries in the past will eventually silt in and become significant wetlands. "Put and take" fishing with a catchable stocking program in these lakes can provide recreational fishing, but someday such lakes will no longer be viable fisheries.

Conservation Pools

For many years anglers have enjoyed fishing the reservoirs in Colorado, but few have known the effort and expense that the Division of Wildlife and water users have made to provide fishing opportunities. Many dams in southwestern Colorado were built by the Bureau of

Reclamation and were designed to maintain minimum conservation pools that provide some excellent fishing. But in some areas such as the Upper Rio Grande the reservoirs were built with private money, and conservation pools were not in anyone's mind when they were built. The Division of Wildlife, as well as ranchers and farmers, have invested great expense and labor that have provided these beautiful places to fish and enjoy. Many of the fisheries in the Upper Rio Grande are fluctuating irrigation reservoirs that do not winterkill. Some of these larger impoundments such as Rio Grande and Continental Reservoirs have produced excellent fishing, but have been drained in dry years, ruining the fishing in both reservoirs and downstream. In 1967 Floyd Getz, Game, Fish and Parks Commissioner, was instrumental in negotiating the first conservation pool agreement with the San Luis Irrigation Company, which owns Rio Grande Reservoir. During the years between 1967 and 1987 Game, Fish and Parks (which became the Division of Wildlife in 1973) negotiated water trades to leave a minimum amount of water for a fisheries pool. Fish stocking rates were based on those minimum pools and provided some of the best fishing in Colorado.

One of the inherent problems with maintaining permanent conservation pools is that water will eventually wear out man-made structures such as the control gates, tunnels, and the dams themselves. These require constant maintenance and sometimes require draining a reservoir completely to make necessary repairs. In those years the fishery can be lost and there is no remedy. These reservoir districts are private and pay for their operations by assessing farmers and ranchers in the San Luis Valley who own the water storage rights. For many decades the farmers maintained these reservoirs and anglers and fishermen dependent businesses enjoyed the fruits of agriculture's investment. Some anglers felt the water users had a moral obligation to maintain those reservoirs for their fishing enjoyment. The irrigation districts' mission, however, was to supply water for agriculture, not fishing recreation. The reservoir operators, facing multi-million dollar repair bills in the 1980s, looked for additional revenue sources to help pay for dam repairs and maintenance. In their view the Division of Wildlife was a likely source. It seemed fair to the water users that anglers who benefit from their reservoirs should share in the cost of running them. The Division of Wildlife and the irrigation districts for

When Rio Grande Reservoir had a conservation pool it offered excellent fishing in a beautiful setting. The Weminuche Wilderness boundary is the high water mark on the south side of the reservoir. Snow capped Pole Mountain is at the upper end of the reservoir. Photo by author

Continental and Rio Grande Reservoirs became deadlocked in their negotiations to have conservation pool agreements. The price and terms of contracts on these two reservoirs could not be agreed upon. Since the negotiations failed, these two reservoirs have periodically drained. Except for drought years they normally are not completely drained.

The Division of Wildlife has conservation pool agreements on hundreds of other reservoirs that provide excellent fishing throughout Colorado. Each agreement has been tailored to meet the needs of water users as well as those of the public. Anglers' dollars pay for these waters and land areas that are open to the general public, often at no cost.

The greatest success the Division of Wildlife has had with conservation pools was through agreements with individuals or small groups of water users who owned water storage rights on small reservoirs in the high country. Their reservoirs often provided habitat for fish and other wildlife, but when the water was drained for irrigation the habitat was lost or severely degraded. For example, in the Upper Rio Grande ranchers built dams on small natural lakes such as Trout, Goose, Shaw, Regan,

A conservation pool in Goose Lake not only stabilizes an excellent high lake fishery, but also enhances a more natural appearance in the Weminuche Wilderness. Photo by author

Poage a and few other lakes. The Division of Wildlife entered into conservation pool agreements with individuals or small groups of ranchers and farmers who own the water rights on these lakes. The Division simply traded water on paper, providing the same volume of water to a landowner's crops, but leaving water in a reservoir. If the Division could not meet its obligation, the water user was free to use his or her water. It has been a win-win agreement for the water user, the Division of Wildlife, and anglers. Stabilized water levels increased the chances of fish not only surviving, but fingerling-sized fish could be stocked that would grow larger because the food base was more stable. An additional benefit was for those lakes in the Weminuche Wilderness that were enhanced aesthetically, because the lakes were left full without the high water mark of a fluctuating reservoir.

Transmountain Water Diversions

It became apparent in the 1970s that the Division of Wildlife was going to need additional water if it was going to be able to accommodate future conservation pools and habitat projects. There were few

The Tabor Diversion, on the summit of Spring Creek Pass, diverts water across the continental divide. It is measured, recorded, and the data relayed via satellite to the Division of Water Resources. The water is traded on paper with ranchers and farmers to maintain conservation pools and wetlands. Photo by author

sources of water available, except for water outside the Rio Grande Basin. Transmountain diversions in Colorado have dewatered several high mountain watersheds by transporting water from the Western Slope to the Eastern Slope. The Division of Wildlife has historically been opposed to transmountain diversions that dewater streams and destroy fisheries. It seems inconsistent for the Division of Wildlife to entertain the idea of owning transmountain diversions, but the purchase of three existing transmountain diversions and changing the beneficial use from agriculture to wildlife has been a forward looking approach to perpetuate wildlife, especially fisheries. These three small transmountain diversions which the Division of Wildlife purchased in the Upper Rio Grande did not dewater drainages on the Gunnison and San Juan sides of the Continental Divide.

The Game and Fish Department bought the Tabor transmountain water diversion in 1961 from Edward G. Lohr of Del Norte. This diversion transports water from a small tributary on Cebolla Creek and diverts it around the mountainside to spill across Spring Creek Pass where Colorado Highway 149 crosses the Continental Divide. But with only one transmountain diversion and Beaver Creek Reservoir for storage, the volume of water was insufficient to meet the goals of improving fisheries in the Upper Rio Grande at that time.

Don La Font and Harley Fuchs built what is now called the Piedra Diversion in 1938 to provide water for their ranches. The

Division of Wildlife purchased this second diversion from Grant Oxley and Sid Klecker in 1970. The diversion is a network of three ditches on the Continental Divide that divert water from the West Fork of the San Juan and East Fork of the Piedra Rivers and dump the water into Red Mountain Creek on the Rio Grande. The water was measured in flumes and documented on recorders which were read each week, and given to the State Water Commissioner. The Division of Wildlife was then credited for that water and could trade it for conservation pools, mitigating wells for wetland maintenance, or anything else within the laws and regulations administered by the Colorado Division of Water Resources. It may sound simple, but the effort to deliver that water has been monumental. The water from these diversions is mainly from snowmelt. The Piedra diversion is about twenty miles west of Wolf Creek Pass, which has the greatest snowfall in Colorado. In the spring the snow is a barrier to water in the ditches and by the time they are free of snow, the snowpack has melted and very little water can be transported across the Divide. Since this diversion is within the edge of the Weminuche Wilderness all maintenance work must be done with hand shovel, wheelbarrow, and sandbags.

In 1971 Howard Spear, Wildlife Technician, and the author found that several years of runoff had filled the ditches with gravel, silt, and rock and little water could be transported. There was too much debris to move by hand so Spear packed dynamite, cement, tools, and a wheelbarrow up to the ditches with Punk Cochran's mules, Jessie and Whiskey. Spear and the author blasted enough of the sand and gravel out of the ditch to make it usable. Spear said, "We just walked into Tomkins Hardware in Creede and bought cases of ditching dynamite, blasting caps, and fuse. No questions asked in those days."

In the spring of 1972 Spear and the author decided they would try to open the ditches earlier to divert more spring runoff. Spear remembers:

> We didn't know what we were getting into. It was June, the snow was melted in the low country, and we thought we could get the water started early. We were young, strong, and enthusiastic, but not experienced with managing water. We hiked and cross-country skied from Ivy Creek Campground nine miles up Red Mountain Creek to the

When diversion ditches washed out, all repair work had to be done by hand. Left to right Ron Desilet, Mike Zgainer, Tom Rauch and Ron Velarde sandbag a washout on the Piedra Diversion in July, 1973. Photo by author

Continental Divide. By the time we got to the Division's one- room cabin near the Continental Divide we were worn out. To lighten the load in our backpacks we had Walt Schuett, our supervisor, and Gordon Saville, the regional pilot, air drop our food at the cabin. When we got there, we found our food scattered near the cabin and one box of food impaled in the tip-top of a spruce tree. We retrieved our food and rested. The snow was so deep that we couldn't even find the ditches. We were there much too early.

A few weeks later when the snow did melt and the head gates could be opened, the water flowed down the ditches only to leak through gopher holes that would wash out the entire ditch if not filled.

Division personnel from the San Luis Valley spent hundreds of man-hours cleaning ditches and filling sandbags. In 1973 Tom Martin, a lanky young man, was the wildlife technician who took on the diversions with great enthusiasm. Coal dust on highways melted snow so why not apply coal dust in the spring to the snow-covered ditches to melt the snow faster? So Gordon Saville, Regional Pilot, dropped bags of coal dust on the diversion ditches. Martin and the author skied up to the diversions and manually scattered the black coal dust. All this effort produced no results.

At 12,000 feet the ditch on the West Fork of the San Juan had a huge snowdrift that prevented water diversion as late as August. Dynamite charges were used in an attempt to blast the snow out of this ditch, but to no avail. Still determined to increase the water for more conservation pools, the next year the Division of Wildlife hired Rex Sheppard, Rich Ormsby, and Allan Czencush to go with Martin and the author to clear one of the ditches. A helicopter flew the crew to the Division of Wildlife's cabin along with provisions and shovels. The ditches were buried under nearly six feet of snow, but we were a determined crew and were fresh. For three days we shoveled snow from the east ditch on the East Fork of the Piedra. After three days of hard work on approximately 1,000 feet of ditch we still hadn't cleared it so that water could flow. After years of very hard work with discouraging results there was a temptation to give up, but the desire of district wildlife managers and biologists to increase conservation pools and improve wildlife habitat was a greater motivation. In 1979 someone came up with the idea of building a pipeline in the Weminuche Wilderness to transport the water across the Continental Divide instead of using open ditches. Nobody accepts the credit or the blame for the idea, but as Tom Martin said:

> It seems logical that any one of us, standing in a ditch at 12,000 feet with a shovel in hand, gasping for air, watching ditch water disappear through a gopher hole or standing in a hand dug trench in a snow drift higher than our heads may have been adequately inspired to come up with the concept of building a pipeline in a wilderness area.
>
> We felt that water would become even more important and valuable and there aren't many ways to increase the water for wildlife such as conservation pools and maintenance of wetlands in the valley floor. We went through quite a process addressing the issues of the Forest Service as they interpreted the 1964 Wilderness Act as it applied to water diversions that predated the Weminuche Wilderness designa-

An Army Chinook helicopter flies a backhoe to the Piedra Diversion on the West Fork of the San Juan River at 12,000 feet. Photo by author

tion. Building such a project would require the use of machinery, a practice specifically prohibited in a wilderness. This would require a special permit. We couldn't pursue getting the legislature to approve funding unless we knew ahead of time that the Forest Service would let us use machinery. The task was too great to be done by hand. We really didn't know if we could even get machinery up to the diversions since the mountains are so steep and rugged.

Regardless of the obstacles, local personnel put together a public information blitz showing what had already been done with the diverted water and what could be done in the future. Concerned ranchers and farmers learned that in fact the Division of Wildlife's projects to date had helped those ranchers and farmers who had cooperated with them. The presentation also demonstrated the problems of managing this isolated water diversion. Almost all of the public supported the proposal. Even those from the San Juan side of the Continental Divide thought the Division was doing the right thing. No one from the San Luis Valley opposed bringing more water into the Rio Grande Basin.

The Piedra transmountain water flows across the continental divide through a buried pipeline to benefit wildlife. The scars of old ditches have now healed and the wilderness character has been restored. Photo by author

The Forest Service even approved of the idea, because the project would cover up the ditch scars in the wilderness and stabilize some wilderness reservoirs.

It took three years for the paper work, planning, and design and legislative approval before the project could get started. Ed Dumph was the wildlife technician in 1979 who was the liaison between the Division of Wildlife and the Forest Service for the project, which started in 1982. Jim Nelson, Division of Wildlife engineer, arranged for the Army to bring in a CH-47 Chinook helicopter as part of a training mission. The Chinook flew a backhoe to the high ditch at 12,000 feet. There was no other way to get the necessary equipment up to the diversion. The state hired a high altitude Llama helicopter to fly the aluminum pipe to the site, some of which was thirty inches in diameter.

The Division of Wildlife invested approximately $300,000 in the pipeline project. The project was started by contractor Dan Counch and finished by Floyd Fredrickson of the Evans Company in August of 1984. There were still some problems with pipe leaks. Ed Dumph said:

> We accepted the project even though there were some pipe leaks. One of the leaks was on the highest diversion. I hiked up there one day and using a backpacker headlight, I crawled down a 24 inch pipe to the first joint and plugged the leak with some sealer. There I was, alone, in a wilderness area at 12,000 feet, down that pipeline and no one knew where I was. If I couldn't have pushed myself backward out of that pipe I was in big trouble. That was really scary, but I got the leak plugged and made it out safely.

Jeff Johnson, Wildlife Technician, said that when the project was finished the Piedra pipeline project solved the problem of snowdrifts blocking the ditches to transport the early runoff, but he still had to make periodic trips to check for and repair any leaks, clean trash racks, and make sure the satellite monitors and recorders were accurate. "It is still a very long and tiring trip," according to Johnson. Anglers and others who enjoyed the benefits of the conservation pools never knew of the weekly horseback rides often in the snow, sleet, wind, rain, lightning, and the backbreaking maintenance that went with each ride.

Wetlands

Wildlife professionals were becoming alarmed at the loss of wetlands in Colorado in the 1970s. Historic wetlands in the San Luis Valley were the low areas that filled with water either from meanders of rivers changing to form sloughs or in many cases artesian water reaching the surface. They became important habitat for ducks, geese, and dozens of species of water birds such as the sandhill cranes. The rare white-faced ibis, and more recently white pelicans have established nesting sites. Like other agricultural areas, farmers found that center-pivot irrigation was more efficient than flood irrigating. During the 1970s many farms went to this form of irrigation and pumping water from the underlying aquifer lowered the water table and dried some historic wetlands. Since the San Luis Valley had some of the most productive and extensive wetlands in southwestern Colorado, Division of Wildlife personnel decided to do something about these losses. The division had already provided some of its surplus water at different times to the Monte Vista National Wildlife Refuge to stabilize some wetlands as well as maintain the wetlands in the Monte Vista State

Wildlife Area, but it was apparent that it would take more water to maintain, let alone increase, wetland habitat.

Just how the Division of Wildlife could get more water for wetlands was a dilemma. The future of wildlife in those areas depended on whether wetlands could be perpetuated. District Wildlife Manager Mike Zgainer talked to several landowners and water users to learn about their water needs and seek some common ground for some water partnerships. The Weminuche Transmountain Diversion could be purchased. Zgainer was instrumental in negotiating the purchase, preparing an Environmental Assessment, and convincing the Joint Budget Committee how important water was to the future of wildlife. The bill passed and the diversion purchased.

Of the three transmountain water diversions owned by the Division of Wildlife the Weminuche diversion produces the greatest volume of water for wildlife in the Rio Grande. This diversion has also challenged the ingenuity of the Division of Wildlife to maintain another diversion in the Weminuche Wilderness. With special permits, helicopters have been used to transport some heavy structures, but all of the work on this diversion including removing silt, sandbagging washouts, and building a concrete diversion box on Rincon La Vaca Creek has been done by backbreaking hand work and by packing materials and workers six miles horseback from the Rio Grande Reservoir.

In the 1990s modern technology changed some aspects of managing these diversions for wildlife. Solar powered transmitters now use satellite technology to relay recordings of water volumes being transported by the diversions to the Division of Water Resources. This saves hundreds of hours for personnel who formerly rode to the three diversions weekly to change water recording charts.

As a result of the Division of Wildlife's effort more than six thousand acres of state owned wetlands are productive and stable in the San Luis Valley. Waterfowl that are hunted as well as thousands of other water birds have benefitted from these projects. Sometimes it takes an "unnatural" approach to perpetuate the "natural." Such has been the Division of Wildlife's strategy to preserving wildlife habitat in the San Luis Valley.

Chapter 9

MANAGING FISH IN TODAY'S WORLD

Until the 1970s, fish management in Colorado consisted primarily of stocking fish, enforcing fishing regulations, and purchasing angler access. With ever increasing pressure on the fisheries, fresh approaches which used science and technology were needed, because the old ways were not keeping up with changing needs. The Division of Wildlife tried to control fish such as the sucker that competed for food and space with trout. The agency experimented with different fishing regulations. In cooperation with business and Trout Unlimited the division sponsored angler education clinics to teach anglers how to catch and release fish rather than killing every fish they could hook. Some anglers were beginning to change some of their priorities. The Division of Wildlife also changed fish stocking policy and fishing regulations to meet the needs of a diverse group of anglers. Attention had been given to threatened and endangered terrestrial wildlife, and now it was time to prevent the decline and even recover the Rio Grande and Colorado River cutthroat trout from its slide toward extinction.

Nemesis—the Western White Sucker

There were several major obstacles to providing consistently good fishing. The western white sucker became one of the major problems of managing the fishery in some reservoirs and lakes in western Colorado. The western white sucker is native to the Arkansas River and the South Platte River drainages. It was likely transported into the San Luis Valley by bait fishermen and later by frequent transfer of fish from private, state, and federal fish hatcheries. Most fish hatcheries utilize river water at some stage in rearing fish, and the tiny sucker fry swim into a hatchery or rearing pond and are stocked out with the trout.

Many attempts have been made to control undesirable fish populations. Rotenone, however, has been the only effective agent used against suckers. In the 1930s rotenone was processed from the derris plant, which grew in the Philippine Islands. Since World War II it has

At Road Canyon reservoir there were very few trout, but thousands of dead suckers and tench piled along the shoreline. Later the reservoir was restocked with rainbow and brook trout and once again became a favorite fishing spot. (September, 1974). Photo by author

been processed from the Barbasco plant, which grows on the eastern slope of the Andes mountains in Peru. Rotenone kills all gill-breathing organisms by coating the gills and preventing the exchange of oxygen and carbon dioxide. The fish die of asphyxiation. Today this suffocant is approved by the Environmental Protection Agency, and applicators must be trained and certified.

One of the most successful applications of rotenone in southwestern Colorado was made at Road Canyon Reservoir in 1974. In 1973 a gill net caught about 300 suckers, 200 tench, and only eleven trout. No wonder the trout fishing was poor! The following year the Division of Wildlife drained the lake to the size of a small pond. Wildlife personnel applied rotenone and people gathered to salvage what some thought would be a tremendous trout population, but to their disappointment and the Division of Wildlife's expectation, there were few trout and innumerable quantities of suckers and tench. Personnel also sprayed rotenone into Upper Road Canyon Reservoir and every puddle and wet spot up to the headwaters.

Great care was taken to restock the lake with fish from sucker-free hatcheries such as the Pitkin hatchery near Gunnison. The Durango hatchery stocked brook trout fingerlings. To this day a sucker has not been seen in Road Canyon Reservoir.

Some private lakes also have sucker problems. For example, private lake owners at Pearl Lakes, Hermit Lakes, S Lazy U, the Wilderness Ranch, and Santa Maria Reservoir have had sucker problems and used rotenone to remove suckers. Suckers seem to return either from being restocked via hatcheries, not all the suckers being killed or swimming upstream.

Most derris projects (projects that use a suffocant to kill fish) have short term results and have to be repeated. North Clear Creek has been treated with rotenone several times. The Division of Wildlife attempted to remove the suckers from North Clear Creek in 1982 when the Pearl Lakes Trout Club, Santa Maria Reservoir Company (which owns the Continental Reservoir dam), and the Division of Wildlife coordinated a derris operation involving dozens of people. Wildlife personnel and volunteers applied rotenone to creeks, seeps, beaver dams, springs, swamps, potholes-every place a sucker could live on both public and private waters. The Continental Reservoir basin had several potholes and each one was sprayed, both from the air and by crews slogging through deep mud. The project went well until a sudden thunderstorm brewed up in the afternoon and almost immediately there was a surge of water going downstream. The neutralizing station that dispersed potassium permanganate was operating as planned just above North Clear Creek Falls. The increased volume and velocity of water exceeded the capacity to neutralize, and the untreated rotenone swept past the neutralizing station and on downstream killing fish in Clear Creek all the way to the Rio Grande. Dead fish floated for many miles and the extent of the loss was really never known, because at the same time the Rio Grande Reservoir was being drained, and there were heavy rainstorms washing silt from that reservoir. The river looked like a chocolate milk shake.

Lloyd Hazzard, Clayton Wetherill, (Area Supervisor,) and the author met with about seventy-five concerned resort owners, business people, and anglers. The tone of the meeting was as somber as a funeral. One seldom sees public officials take responsibility for events that go wrong, but what happened next was just that.

Hazzard began the meeting:

> Ladies and gentlemen I can't tell you how sorry I am that this accident has happened. If you are looking for the person responsible, you are looking at him. I want to explain to you what we were trying to accomplish, what we think went wrong, and what actions we are preparing to take to mitigate this tragedy.

Lloyd Hazzard talks to the crowd of concerned resort and business owners at the Bristol Inn Restaurant following the accidental fish kill on Clear Creek. The author and Clayton Wetherill are seated. Photo from Division of Wildlife, Meg Gallagher photographer

You will remember that a few years ago you wouldn't send any of your guests to Continental Reservoir, because it was full of suckers. After we derrised the reservoir, established a conservation pool with the Santa Maria Reservoir Company, and restocked the reservoir, it became one of the best trout fisheries in Colorado. Well, the suckers came back into that drainage. We received notice that the reservoir was going to be drained for repairs and it offered us an excellent opportunity to derris the drainage again and start all over. We were looking forward again to maintaining the excellent fishing that you all have enjoyed.

It was a difficult time for these men who were dedicated to improving the fishing for these people, who were justifiably angry over the devastating accident. But after the meeting Floyd Freemon, former owner of the Freemon Ranch, went up to the author and put a big hand on his shoulder and said, "'We're mad today, but we'll get over it. Don't you give up trying to improve the fishing." Others said that they were upset, but understood the goal. They asked only that if it ever had to be done again they hoped that it could be done safely. Some said that if the Division of Wildlife needed help in the future to just let them know how they could be of assistance. The forgiveness and encouragement stimulated greater cooperation and understanding in the days and years to come.

It did take time for feelings to heal, but the process started the next day when about a dozen fish trucks started restocking fish. Hazzard diverted fish from other plants to the Rio Grande and Clear Creek. Ken Ellison, owner of the Freemon Ranch, packed fish by horseback up Clear Creek. Volunteers helped restock every place that had been affected. After the restocking effort was completed, one landowner was heard to say, "Fishing hasn't been this good in a long time." Yet, many large and wild fish were lost and it took several years for a full recovery.

The Division of Wildlife, in cooperation with the Hermit Lakes Club on South Clear Creek, coordinated rotenone projects in 1968, 1978, and 1998. In each attempt all the lakes were drained as low as possible to reduce the surface area and water volume. Teams applied the suffocant to every place that could be reached. Unfortunately, with the swampy conditions around the lakes, it has been impossible to kill all the suckers. A rotenone project probably will be needed periodically to control the sucker population. And yet, according to Dean Prentice, veteran sucker killer:

> This year (1998) I think we got all of them. This time we used a commercial crop duster to spray rotenone on places that couldn't be reached on foot. We've learned to pay very close attention to how we do this now. We've set gill nets and they came up empty. We've also built sucker barriers to keep them out of our lakes.

A year later no suckers had been seen in the drainage above South Clear Creek Falls.

Changes in Fish Stocking

The Division of Wildlife made more improvements to its fish distribution in the 1970s when Lloyd Hazzard, Regional Fish Biologist, modernized the hatchery truck fleet in southwestern Colorado. The old steel tanks on the standard trucks were gradually replaced with new, insulated fiberglass tanks that were lighter in weight and could carry more fish farther and deliver them in better condition. There was some initial opposition to these new trucks by veteran fish culturists who were slow to accept change. Some of the public were so used to the old trucks that they thought the state had stopped stocking, because they no longer saw the old steel tanks on trucks with the noisy gasoline pony motors recirculating water. Such improvements have resulted in greater efficiency and cost savings that went unnoticed by most citizens.

Punk Cochran, like most Wildlife Conservation Officers, regularly met the fish trucks and helped distribute fish (May, 1967). Photo by author

In the early 1970s the Game, Fish and Parks Department was having problems with fish diseases in several of its hatcheries. The state needed a consistent source of healthy rainbow eggs. The division leased the Hartman Fish Hatchery (formerly the Hosselkus hatchery) from Mervin Hartman in 1973 in order to have a brood fish station as a secure source of rainbow eggs. Bob Little was assigned as superintendent with responsibilities for the operation. The state spawned about two million rainbow eggs annually. Most of the fish eggs were sent to other state hatcheries. Enough were hatched at the unit to maintain the brood stock. By 1978 the state no longer needed the production of this unit, and the lease was terminated.

In 1968 Dave Lemons, Regional Fish Biologist, initiated a high lake inventory project in southwestern Colorado. No one had ever compiled basic information about all the high lakes. The lakes were stocked, but little was known about the fish populations, lake conditions or if the lakes were being properly managed. Lloyd Hazzard, the field biologist at that time, supervised a crew to survey every lake in southwestern Colorado. They gathered basic physical, chemical, and biological data about each lake. This information became the basis for determining species and stocking rates in the high country lakes. Eventually all lakes and reservoirs were similarly surveyed.

Fish stocking continued to change after the 1970s. It was becoming apparent that the state stocking program was not satisfactory. Earlier the hatcheries stocked mostly fingerling-sized trout in lakes, streams and beaver ponds. Stocking six-to seven-inch fish from stream side ponds and hatcheries started in 1927 when the Game and Fish Department acquired its first aerated fish truck. Under Director Rolland Parvin, the Game and Fish Department started stocking catchable fish (over eight inches) when an over abundance of spawning male rainbow trout occurred at Parvin Lake in 1930. These larger brood fish were stocked in several streams in northern Colorado, resulting in great publicity in local newspapers. The entire procedure was made into a film entitled "Chasing Rainbows" and released for public consumption. The catchable program was expanded under C. N. Feast in the 1940s and gave birth to the "monster" that many anglers loved and most of the resort-based businesses came to expect. The public demanded more catchable-sized fish. The state responded by building efficient raceways and retaining ponds at its hatcheries that could handle larger fish, even though the fish were much more expensive to stock.

Colorado then made some changes. The division decided to manage some waters as "put and grow" by stocking fingerlings, some "put and take" by stocking catchables, and others designated as "wild" which would not be stocked. The Upper Rio Grande was well suited to this management approach. Some reservoirs and lakes such as Rio Grande, Continental, and Road Canyon Reservoirs could be stocked every year with two-to-four-inch trout. The next year they would be nice fat catchable trout. Others were stocked with catchable-sized fish most of which would be caught by the end of the summer. Some waters, such as the Rio Grande itself, have stream sections where the habitat sustains wild populations. Stocking domesticated fish on top of these wild populations increased the competition for space and food. It is true that the wild fish may have been more difficult to catch, but it was getting to the point that if the state stocked fish that were any dumber, they wouldn't have known how to swim. The new guidelines still provided some flexibility that allows the state to stock Colorado River rainbow fingerling which were from wild stock.

Lloyd Hazzard found that the Snake River cutthroat trout had some characteristics that would enhance fishing opportunity. This subspecies of cutthroat had more efficient gill rakers that allowed it to more efficiently utilize zooplankton (macroscopic organisms) in its diet. The fish could be managed to spawn in the fall and provided the opportunity to stock large fingerlings the following spring. Hazzard and Lawrence Lillipop, Fish Culturist at the Pitkin hatchery, hauled the first Snake River cutthroats to Colorado from Montana and established a brood fish population at the Pitkin and Roaring Judy hatcheries near Gunnison. These fish were stocked in Continental Reservoir and many of the high lakes of southwestern Colorado and have provided excellent fishing.

Stream Improvement

About the time the state changed stocking policies it was also evaluating stream habitat. Determining pool-riffle ratios and evaluating the carrying capacity of streams was an initial step to many stream improvement projects in Colorado. By manipulating boulders into certain configurations, the carrying capacity of the stream for fish could be improved. Some stretches such as the Rio Grande in the Coller State Wildlife Area are mostly flat water at places where the river is wide and shallow, with few deep pools to sustain a good fish. In 1975 Mike Zgainer initiated a stream improvement project on three miles of river. Rick Sherman, Regional Habitat Biologist, obtained the federal permits for making stream improvements and planned the boulder placements. The project only cost about $5,000, because Rio Grande County was hired to stockpile boulders and use their large rubber tired front-end loader to place the rocks. Rock placements with various configurations added structure to the river and increased the number and depth of pools in that section of river. Later, another project completed the stream improvement work on the Coller State Wildlife Area. The brown trout population increased, but improving the structure was not enough. Fishermen were catching the fish faster than the stream could grow them.

The Division of Wildlife, in cooperation with the Forest Service and Trout Unlimited, rebuilt several small streams in southwestern

A fly fisherman fishes pools created by what appear to be natural boulders at the Coller State Wildlife area. Photo by author

Colorado. One such project was undertaken on North Clear Creek in 1985. This beautiful meandering stream above North Clear Creek Falls was very popular with anglers, but the riparian area was heavily grazed by cattle and the stream banks were constantly collapsing so that there were very few pools that would sustain fish. The stream flow is reduced every fall when the Continental Reservoir begins to store water and that further reduces habitat.

The Division of Wildlife, Forest Service, and Trout Unlimited volunteers built a variety of structures, dams, baffles, rock placements, and willow plantings to stabilize the stream and improve the fish habitat. Bill Babcock, Research Biologist, evaluated the water quality before construction began. A contractor placed large boulders and moved mature willow bushes to strategic locations along the stream. The Forest Service furnished personnel to prepare the logs for placement. Trout Unlimited volunteers planted willows along the stream-side to stabilize the banks. As a result the pool-riffle ratio was improved making it a better fishing spot. Continued overgrazing by cattle in the riparian area however, has limited the full potential of the project.

A number of private landowners in southwestern Colorado have seen the benefits of managing their stream sections and protecting riparian areas. For example, the La Garita and 4UR ranches hired private contractors to reconstruct Bellows and Goose Creeks. Years of cattle grazing had destroyed much of the riparian structure along these beautiful streams. After reconstruction these stream sections provided much improved fishing for guests of these ranches.

Water Pollution

Historic and ongoing stream pollution has been a limiting factor in many Colorado rivers and streams. In 1971 Barry Nehring, a student at Colorado State University, wrote his Masters Thesis on the Rio Grande, studying the accumulation of lead, zinc, copper, and silver in Mayfly and Stonefly insects. High levels of heavy metals resulting from decades of mining and milling activity, left deposits of these elements in toxic levels throughout the floodplain of Willow Creek between Creede and the Rio Grande. Even low levels of these elements are toxic to trout. It was unknown how toxic they were to insects in the food chain.

Using the Creede hatchery facility, Nehring monitored the effects of various concentrations of polluted Willow Creek water on insects and small trout. He studied the effects of these metals on rainbow, brook, brown, and cutthroat as well as on the insects. Nehring found that levels as low as six parts per million could kill trout in a few hours. He found that these elements plated out on the exoskeleton of the insects and was shed, so they were less susceptible to the pollution than anticipated. The trout that feed on these insects did not show excessive levels of heavy metals in their flesh.

A sudden surge of toxic water however, was an altogether different event, as was documented during the course of the study when a flash flood washed out a drainage ditch below the Emperius Mill and dumped mill effluent into Willow Creek and the Rio Grande River. One of Nehring's experiments was under way when the effluent hit the hatchery intake. The fish inside immediately started dying. This was one of the best documented pollution cases in Colorado. The Colorado Water Quality Control Commission met to hear the case against the Emperius Mining Company. Nehring and the author testified in that hearing. The commission took no action at that hearing, but because of pressure from Trout Unlimited and others, the commission had a second formal hearing in which Nehring's testimony was convincing enough for a "Cease and Desist" order to be issued to Emperius. A previous spill had occurred in 1963. Both resulted in massive fish kills between Creede and South Fork. In both cases no mitigation was required.

Most fisherman don't even dream of catching a large fish like Robert Zabrowski's eighteen pound, thirty-four inch German Brown that he caught from the Rio Grande in 1972. Restrictive fishing regulations on some stretches of rivers has improved the opportunity to catch wild and large fish once again. Photo by author

Science-Based Fishing Regulations

The quality of fishing continued to decline and old solutions were not keeping up. Some private landowners have initiated their own scientific river studies in order to make good decisions about what restrictions or renovations would be the best investment for their properties. In 1977 the Wason Ranch hired Leroy Fyock, a Creede native son who was finishing his fisheries studies at Colorado State University, to study water quality and the fish population. As a result of Fyock's study, the Wason Ranch required their guests to release all fish less than fourteen inches. By 1982 it was evident that the ranch-imposed restriction was not increasing the number of larger fish. Barry Nehring, Fish Research Biologist, then began an intensive evaluation of special regulations on the Rio Grande. Nehring recommended to the Wason Ranch that they allow anglers to keep two fish less than fourteen inches and allow the larger fish to live and reproduce. By 1985 the number of larger fish had doubled. The knowledge gained on the Wason Ranch and Coller sections was valuable to the

Barry Nehring's fish crew floating the Rio Grande. An electrical generator supplies the mild "shock" that stuns fish when an electrode is thrown into the river. Fish are attracted to the electrode and are netted. Fish are then measured, weighed, and safely released. Photo by author

understanding of how the proper bag limit restrictions could improve fishing.

Nehring led a team of research biologists studying other rivers in Colorado and evaluating various special fishing regulations. As a result regulations were modified on many streams so as to be compatible with the biological potential of a given river section.

Even though restrictive regulations and the stocking of wild rainbow trout on rivers such as the Colorado was improving the fishing, something had to be done to change angler attitudes, prior to having special bag limit restrictions. In 1985 the author began contacting businesses, landowners, resort owners, and many fishermen to explore the possibility of starting an educational program and voluntary bag limit restrictions on the Rio Grande that would lay a new foundation for improving the fishing. The community was supportive of such an effort. Meg Gallagher, Regional Information Specialist, developed a brochure to explain the goals and details of a voluntary program which would restrict the kill of rainbows more than twelve inches by releasing the larger fish to fight another day and perhaps even spawn. The following year nearly seventy-five percent of the fishermen contacted were voluntarily complying with the restriction.

At this same time, Nehring began stocking fingerling Colorado River rainbows to boost the Rio Grande's stock of wild rainbows. By 1987 the rainbow population in the Rio Grande had increased dramatically, and most fishermen were pleased with the results. The Wildlife Commission made a regulation that applied to three miles of river above Marshal Park Campground in 1988 to restrict the bag and size limit. Such restrictions were not acceptable to some fishermen and resort owners. Through Nehring's research project he determined that special restrictions in some river sections would not produce larger fish, because the habitat was inadequate. In those sections "put and take" stocking of catchable rainbows continued.

According to Jay Sarason, District Wildlife Manager, "There was opposition to mandatory bag limit restrictions. A lot of grandparents have told stories of the fabulous fishing of the past and want their grandchildren to make those same memories for their children."

The rivers of Colorado cannot sustain the memories of the past. This is the reason that the new generation needs have realistic dreams and make new memories. Education can change some people's minds and their dreams. The Division of Wildlife started an educational program in 1985 to help anglers become better fishermen and to encourage a new attitude toward fishing in general. Several Trout Unlimited chapters and individuals came from far and wide to sponsor fishing clinics at campgrounds and the Creede hatchery in order to teach anglers how to tie flies, cast a fly rod, and carefully release unharmed fish. If anglers become more educated and skilled in the sport, then perhaps they will get more out of their fishing experience than just seeing how many or how large a fish they can kill.

Soon there were restrictive fishing regulations on a number of rivers in Colorado. With the increase of restrictive regulations came the problem of patrolling the rivers for violators. Idaho Conservation Officers had used kayaks to patrol some of their wilderness rivers, and so the idea was born for Dave Kenvin, District Wildlife Manager in South Fork and the author to learn how to kayak. The state would not furnish kayaks, equipment, or training, but the San Luis Valley Chap-

A young mother teaches the art of fly casting to her son at a fishing clinic at 30 Mile Campground. (August, 1985). Photo by author

ter of Trout Unlimited raised the money and purchased all the necessary gear. "Buck" Stroh, a business teacher in Creede and an expert kayaker, volunteered to teach them how to kayak. After three hours in the Monte Vista swimming pool learning to do the "Eskimo roll" (half of that time being underwater), Kenvin successfully rolled upright and with chlorine-reddened eyes declared, "The first guy we catch without a license is going to be in very deep . . . trouble."

The following summer many a fisherman was surprised when Kenvin and the author paddled through a rapid to a surprised angler. "We're District Wildlife Managers of the Division of Wildlife; we'd like to check your fishing license and fish please." "You gotta be kidding!" Some thought it unfair that game wardens would be so "sneaky." Others felt their "private" world was violated by these floating officers. Still others were excited, "I've fished up here for fifty years and never been checked-keep up the good work!" Those who had fished in the rugged Box Canyon below Rio Grande Reservoir had seldom if ever been checked. As one violator said, "Officer, you really earned this one and I don't mind paying the fine. I just can't believe anyone would be crazy enough to patrol this canyon in a kayak."

The author (foreground) and Dave Kenvin, (in 1985) patrolled the "rock garden" between Cottonwood Cove and Blue Creek. Other officers soon learned how to kayak and before long the Division of Wildlife's kayaking navy was patrolling most of the rivers in southwestern Colorado. Photo from Division of Wildlife, Meg Gallagher photographer

Angler Access

Angler access to some rivers changed when a Grand County court case stirred up the issue of access to Colorado's rivers. For many decades landowners had filed trespass charges against fishermen who crossed private lands without permission to fish in their rivers. As a result of a 1979 trespass case, Attorney General Duane Woodard issued his opinion in 1983 that one who floats the river without touching bank or bottom does not commit a criminal trespass. Woodard's opinion has not been challenged to this date. The rafting business picked up substantially following the issuance of that opinion. Those ranches which catered to fishermen were the most disturbed by this opinion, because it allowed the public to float and fish through previously private waters that in many cases were privately stocked. The solitude of their guests was shattered by the noise of thrill-seeking rafters. Resort owners on major rivers in Colorado wanted protection from the general public. Some anglers thought they had discovered a trout fishing bonanza by drift fishing through private ranches where ranch-imposed regulations had greatly improved the fishing. The conflict continues to this day.

Whirling Disease

The greatest natural disaster that invaded most of the major watersheds in Colorado was in 1988 in the form of the Whirling Disease parasite. This parasite was imported into the United States from Europe in the 1950s. It found its way west by means of private and public fish stocking. It can also be transported by waterfowl, hauling water, or even the transfer of mud from one drainage to another. Even fishing gear that is used in different waters can carry the parasite. The parasite was discovered in 1988 in the Rio Grande and by 1993 the rainbow population in the Rio Grande was reduced by ninety-five percent. The parasite in one stage is the size of a tiny dot and is translucent so it isn't seen by the naked eye. This life stage floats freely in a river, attaches itself to a trout, and attacks the cartilage of small trout causing deformation of the forming skeleton. When the fish becomes excited, either in feeding or fleeing danger, it swims in circles—thus the name "Whirling Disease." A fish in this young stage is easy prey for larger predator trout. The Division of Wildlife, in cooperation with other states, began studying the dilemma and sharing information to learn how to manage fisheries with Whirling Disease. At this time it is impossible to rid river basins of the parasite. Barry Nehring has become nationally recognized for his research of Whirling Disease. The Division of Wildlife is sterilizing and remodeling hatcheries such as Roaring Judy near Gunnison to produce more Whirling Disease-free fish. The process is very expensive, and it will take perhaps many years for fish biologists to learn how to manage fisheries where Whirling Disease is present. The Durango fish hatchery was certified Whirling Disease free in the summer of 1999 and can stock its fish in waters where the parasite has not been found.

Saving Threatened Species

By 1973 the Rio Grande cutthroat trout had declined to the point where the Colorado Wildlife Commission designated it as a threatened species in Colorado. The Rio Grande cutthroat differs from greenback and Colorado River cutthroat trout by having fewer scales and by the irregular shape of spots on the narrow part of the tail. Rio Grande cutthroat trout develops colors like those of greenback and Colorado River cutthroat trout-rich reds, oranges, and golden yellows, but often not as intense as in other subspecies.

The Division of Wildlife conducted stream and lake surveys and collected fish that were considered genetically pure Rio Grande cutthroats. Only tiny Bennett Creek in the Upper Rio Grande still contained a remnant population of pure Rio Grande cutthroats. There were several reasons for the extirpation of this subspecies of the cutthroat. The cutthroat is the easiest of the trouts to catch so was consequently removed quite rapidly. They also had a genetic propensity to move downstream and were lost in increasing numbers in irrigation ditches. Fall-spawning fish such as the brook are more successful and tend to stay in one location and could out-compete the cutthroat. As more irrigation ditches were built, the return flows to streams caused increased water temperatures to levels favoring more tolerant species. Introduction of other subspecies of cutthroat and rainbow resulted in hybridization and loss of the genetic purity of the native cutthroat. Lastly the cutthroat were more difficult to rear in a hatchery, hence early culturists favored the other trout. One hundred years of stocking fish throughout the Upper Rio Grande had for all practical purposes destroyed this subspecies of cutthroat trout.

Ron Velarde, Regional Manager for western Colorado, said:

> When I was a District Wildlife Manager in Costilla County I was aware that there were Rio Grande cutthroats on the Forbes-Trinchera ranch. I just felt that we should do whatever it took to see to it that this important species survived, not only on private land but public land as well, so that all could enjoy this beautiful fish. We did get initial resistance from the Forest Service, but Lloyd Hazzard and I wrote the Environmental Assessments for the Forest Service and got approval to use rotenone to clean several streams of all fish so that we could establish the Rio Grande cutthroat. We used the media to gain support for our project. Our agreement with Errol Ryland, the Trinchera manager, was to take all the Rio Grande cutthroats we could to restock other streams and then rotenone Trinchera creek and build a fish barrier to prevent migration of non-native fish up into that basin. We restocked Trinchera Creek as well to maintain the Rio Grande cutthroat in that drainage.

> We built fish barriers on Jim and Torsito Creeks west of La Jara Reservoir and derrised those drainages and Dick Weldon packed Rio Grande cutthroat in horseback to get them established. Once we did these drainages, other District Wildlife Managers started their own recovery programs. Lloyd Hazzard was our leader and mentor. We didn't have a budget or big plan. We just did what had to be done.

Velarde's project became the model for successful recovery of the Rio Grande cutthroat in the San Luis Valley. In 1979 Mike Zgainer, District Wildlife Manager, started a Rio Grande cutthroat recovery project on the San Francisco Lakes and that drainage south of Del Norte. Brood stock projects were created at Haypress Lake and the Fish Research Hatchery in Fort Collins. Fingerling cutthroat cultured from these brood fish have been stocked in many wilderness lakes and streams in the Upper Rio Grande as well as other drainages in the Sangre de Cristo Mountain Range. Many of the alpine lakes of these areas do not have natural reproduction and artificial stocking sustains the fishery. These lakes are being stocked and in a few years the existing populations of trout will be replaced by the Rio Grande cutthroats. These beautiful fish have been stocked in Crystal, Goose, Lost Trail and West Lost Trail Creeks, the Rio Grande upriver from Lost Trail Creek, Big and Little Ruby Lake, Squaw Creek, Trout Lake, the Ute Lakes, and Weminuche Creek. Dick Weldon, former District Wildlife Manager in La Jara, was instrumental in reintroduction efforts of Rio Grande cutthroat on several waters in the Conejos River tributaries and in the South San Juan Wilderness Area. Bert Widhalm, former District Wildlife Manager in Saguache, reestablished the Rio Grande cutthroat in several small streams in the Saguache district. With the true native cutthroat becoming self-sustaining in much of its former habitat the Wildlife Commission changed its status from "threatened" to a "species of special concern" in 1984. There are still threats to the Rio Grande cutthroat including Whirling Disease, hybridization, competition with non-native fish, degrading of habitat, and fishing. It will take a continuous effort to maintain the Rio Grande cutthroat, but careful management and citizen awareness will help to ensure its survival.

The Colorado River cutthroat is a species of special concern in Colorado. It is the native trout in the Colorado River including the

San Juan and Gunnison basins. In southwestern Colorado, fourteen wild populations have been found in the San Juan Basin and five in the Gunnison. Some streams have already been reclaimed and more work is planned to establish brood fish populations. The reclamation effort is similar to the Rio Grande cutthroat recovery program.

Another fish on the state's endangered species list is the Rio Grande sucker. The Division of Wildlife has a recovery program for this species, but the project is concentrated in the San Luis Valley. The Division of Wildlife is constructing the Native Aquatic Species Restoration Facility southwest of Alamosa, Colorado that will be strictly dedicated to perpetuating native species such as the Rio Grande sucker. Some sportsmen disagree with bringing back a species that they believe the Division of Wildlife has tried to eradicate, but this species is not the prolific and competitive western white sucker.

The Division of Wildlife has a responsibility not only to sportsmen, who pay most of the bills with their license dollars, but also to all species of wildlife in Colorado, most of which are non-game. Sportsmen's license money has paid for the majority of Colorado's non-game, endangered, and threatened species programs. The Colorado income tax check-off program has contributed substantial amounts as well. The Colorado Lottery has potential to become the major funding source for future management of these animals, birds, and fish.

Chapter 10

MANAGING TODAY'S WILDLIFE HERITAGE

When the Colorado Legislature changed the name of the Game, Fish and Parks Department to the Division of Wildlife in 1972 it initiated a transition from being a "hook and bullet club" to an agency that was given responsibility to perpetuate all wildlife species. The title of "Wildlife Conservation Officer" was changed to "District Wildlife Manager" to reflect that the position was responsible for more than law enforcement, but included the management and protection of all wildlife within a geographical district.

In 1973 Congress passed the Endangered Species Act to prevent additional wildlife from becoming extinct and where possible to recover those species that were in danger of becoming extinct. The endangered species that had been extirpated from southwestern Colorado were the grizzly, river otter, wolverine, black-footed ferret, and peregrine falcon. The bald eagle was rarely seen except in the winter in the San Luis Valley and briefly along other river corridors in the Southwest. They had not been nesting in Colorado for many years.

Wildlife scientists have learned a lot about wildlife, but the animals remain wild and still unpredictable. Biologists have applied scientific principles to refine the art of wildlife management and in most cases have been successful. Most species are better off at the end of the twentieth century than they were fifty years ago. Some have been lost and may never return to the San Juans, but those decisions to reintroduce animals such as the wolf, grizzly, and wolverine will be made in the future. The most important land use decisions affect the remaining wild places.

The Grizzly

The grizzly was given protection under the Endangered Species Act in 1973. The Division of Wildlife had searched for grizzlies for over two decades. Ed Wiseman, an outfitter from Hooper, Colorado, was attacked and mauled by a grizzly while guiding an archery hunter in 1979 in the South San Juan Wilderness near Platoro, Colorado, proving that at least one grizzly remained in the San Juans. Wiseman killed the bear in self-defense. This incident was a springboard for Tom Beck, Division of Wildlife researcher, to initiate yet another search for grizzlies in the southern San Juans. No grizzlies were ever located as a result of his study. Other biologists and citizen groups have searched for grizzlies in the San Juans, but no recent evidence of their existence has been documented. Suspicion of their existence remains, but the grizzly seems to be a ghost in the San Juans that some people hope will reappear. The Colorado Wildlife Commission passed a resolution in 1978 against the reintroduction of grizzly or wolf. Agricultural interests were strongly opposed to such reintroductions. Even as large as the San Juans appear on a map, the habitat is now fragmented with roads, subdivisions, ski areas, golf courses, clearcuts and other developments where reintroduction of the grizzly would create problems which people who live near the area are not prepared to accept. The ecosystem of southwestern Colorado has changed and whether or not such large a predator would survive in today's environment is questionable.

Rocky Mountain Goat

Some non-native species that have been introduced in Colorado have found a niche and have become firmly established. The Rocky Mountain goat was not native to Colorado, but there is suitable habitat for this white-coated denizen of the crags. The Game and Fish Department had been introducing the mountain goat since 1947 and had been very successful in getting it established in the Collegiate Range and on Mount Evans. In the early 1960s there had been a complete die off of bighorn sheep in the Lake City area. Jim Houston, Wildlife Conservation Officer in Lake City, felt that there was an ecological niche in that area that mountain goats could fill,.so he requested a goat transplant. In 1964 four billies and six nannies were brought in from South Dakota and released on Mill Creek. Those animals gradually disappeared.

In 1966 Wayne Knisley, Wildlife Conservation Officer from Durango, was with a party of archers in the Beartown area of the Upper Rio Grande when one of the hunters saw three goats. In 1969 a hiker reported finding the carcass of a tagged mountain goat from the

Lake City transplant in Chicago Basin of the Needle Mountains. With this discovery Bill Rutherford, a wildlife research biologist from Fort Collins, and regional biologist Errol Ryland flew the San Juans in a helicopter to follow up on the report and found a small herd of the goats in Chicago Basin.

Since the mountain goats were firmly established in that area, the department made a transplant in 1971 with mountain goats trapped in British Columbia. It was expensive to capture and transport mountain goats from so great a distance, so in 1973 a third transplant was made from the growing mountain goat population in the Collegiate Range west of Buena Vista, Colorado. The herd increased to the point that there have been limited hunting licenses issued to hunters, and the population continues to thrive. The only conflict with humans occurred when back packers began feeding the goats and some became so tame that they could be hand fed by anyone offering a Twinkie. An education program to encourage hikers to not feed the wild animals has had a positive effect to reduce this conflict.

None of the mountain goats stayed in the Lake City area. In 1988, however, two bighorn sheep hunters reported seeing three mountain goats on Stewart Peak in the La Garita Wilderness. They also observed two poachers killing one of the goats. District Wildlife Manager Phil Mason, the author, and Gunnison Supervisor Jim Houston caught up with the culprits, and they were taken before Mineral County Judge Robert Wardell and fined $1,000. The very expensive goat meat was donated to the Mineral County Senior Citizen hot lunch program. The meal was not greatly appreciated even though the cooks did all they could to tenderize the meat. It was tougher than a boot and some say it took an electric knife to cut the gravy.

River Otter

River otters were probably never abundant but they were native to many river systems in Colorado. They once lived in the San Juans, but were trapped out by the early trappers. Dave Langlois was the first Division of Wildlife non-game biologist in southwestern Colorado to begin surveys for those species that had been listed as endangered and threatened. District Wildlife Managers suggested rivers that could be candidates for river otter reintroduction. River otter require some year-around open water and a sufficient fish population to sustain them. No drainages were found in the Upper Rio Grande that would support river otter. Where there was habitat there was human presence and the probability of a successful transplant was poor.

Tom Beck, research biologist, led the otter reintroduction project and for several years monitored the successful reintroductions that were made in the Piedra and Dolores rivers and in the Black Canyon of the Gunnison. According to Scott Wait, wildlife biologist for the San Juan Basin, "The river otter reintroduction has been a real success story. In 1999 otters were reported in nearly every major drainage in the San Juan Basin as well as the Dolores River." River otter continue to live in the Black Canyon as well. There are other drainages where river otter have been successful, but many of their natural habitats have been destroyed.

Eagles and Falcons

For many years the sighting of a bald eagle was so rare that citizens would report such an event. Eagles migrated from the north to winter in southern Colorado, especially in the San Luis Valley, but they returned northward to nest each spring. There were no known nesting eagles in Colorado. In the 1970s field personnel and wildlife students looked for bald eagle nests throughout the state and found none. The pesticide DDT was responsible for the decline of many raptor populations such as the peregrine and bald eagle all across North America. The pesticide was banned, but it breaks down very slowly. Bald eagle recovery has been slow, but around 1990 a pair of bald eagles established a nest at the Santa Maria Reservoir. Tammy (Fox) Spezze, District Wildlife Manager, said, "I watched the nest, but never saw any fledglings come off." In the spring of 1999 this pair did raise a single fledgling. Even farther south there were three active bald eagle nests in the San Juan Basin by the end of the 1990s.

In the 1990s more than a hundred bald eagles have been gathering every fall on the East River north of Gunnison to feed on migrating kokanee salmon. Some of those eagles spend the winter in the Gunnison Valley. Between 100 and 150 eagles have been wintering in the San Luis Valley. Bald eagles have increased and have been removed from the endangered species list.

There were always a few remaining active peregrine falcon nesting sites in southwestern Colorado, on the Conejos river, Wolf Creek Pass,

Chimney Rock and the Hermosa Cliffs north of Durango. Several observations were made of peregrines in the Upper Rio Grande as late as 1950, but they disappeared. With the assistance of the Peregrine Fund, the Division of Wildlife began an extensive retintroduction program in Colorado. This fastest flying falcon was successfully reintroduced to the Durango and Pagosa Springs area. By the end of the 1990s more than twenty pairs of peregrines had active nests in the San Juan Basin. About six nesting pairs were in the San Luis Valley. Peregrines recovered sufficiently in Colorado that they have been removed from the endangered species list.

The Boreal Toad

The boreal toad, which is an endangered species, was found in the Upper Rio Grande. John Alves, Area Fish Biologist, said that boreal toads have been found in Jumper Creek, Love Lake, Red Mountain Creek, Workman Creek, Trout Creek, Regan Lake, and at the Wetherill Ranch along the Rio Grande. The boreal toad, however, is rare in the San Juan Mountains of Colorado, and its current status has yet to be determined. Such species are considered important indicators of an ecosystem's health and need to be monitored. Historically, most attention has been paid to those species of wildlife that have economic or recreational value. Amphibians and reptiles are nongame wildlife that are receiving more attention from the Division of Wildlife.

Wolverine

The San Juan mountains are on the southernmost edge of the wolverine range in North America. The last record of a wolverine killed in the San Juans was about 1910 in Antelope Park. Since then there have been a number of unconfirmed sightings. Bill McKee, caretaker at Rio Grande Reservoir, reported seeing a wolverine in 1968 northwest of Weminuche Pass while he was archery hunting. He was within one hundred yards and watched it through binoculars. Mark DeGregario, a temporary Forest Service employee, saw a wolverine in Antelope Park one summer evening in 1978:

> I came around a corner just above the Park Corrals on Colorado Highway 149 when this animal ran down onto the road and started running along side of me. It was definitely a wolverine. It paralleled

me for a while and I stopped and it ran across the road and into the willows toward the river. I didn't report it at the time, because I didn't think anyone would believe me.

Dave Kenvin, Terrestrial Biologist for the San Luis Valley, conducted an intensive search for wolverine between 1994 and 1997 in the San Juans. That search didn't produce any sign or sightings of wolverine. Much of the high country of the Weminuche Wilderness is suitable wolverine habitat. The reintroduction of small predators such as the wolverine would have a high probability of success. The Colorado Wildlife Commission supports future wolverine reintroduction into the Weminuche. However, in 1999 the Colorado Legislature passed a law that takes decisions to reintroduce endangered wildlife away from the Wildlife Commission and all such reintroductions must be approved by the legislature.

Lynx

The last Canadian lynx was seen in Colorado in 1973 on Vail Mountain. Lynx has been on the Colorado endangered species list since 1973. In 1996 the Wildlife Commission directed the Division of Wildlife to investigate the possibility of reintroducing lynx to Colorado. Although there were few verified sightings of lynx in the San Juans, this mountain range does have significant lynx habitat and prey base. A United States Fish and Wildlife supervisor reported catching a lynx in 1941 on Cochetopa Pass. Old timers talked about the existence of lynx cats, but careful examination of old photographs show they were most likely large bobcats. Lynx are larger than a bobcat with big paws that aid them in traveling in the snow. The snowshoe hare is its major prey, but it will eat whatever it can catch.

Dave Kenvin, area biologist, along with District Wildlife Managers Jerry Pacheco and Brent Woodward made snowshoe hare surveys in the Upper Rio Grande, while biologist Scott Wait and District Wildlife Managers Glen Eyre, Mike Reid, and Cary Carron inventoried the prey base in the San Juan Basin to determine if it was sufficient. Their surveys indicated there was a sufficient prey base to sustain lynx in the San Juans. The Division of Wildlife held public meetings in the San Luis Valley and Pagosa Springs to determine if there was public support for a reintroduction of lynx. There was general support, except for opposi-

Biologists Dave Kenvin on the left and Jim Olterman release the first Canadian lynx near Humphreys Lake on Goose Creek at the edge of the Weminuche Wilderness. (February, 1999). Photo by Russ Smythe, Montrose Daily Press

tion from the livestock interests. As one rancher said, "My great-granddad, my granddad, my dad, and I have tried to exterminate all these bad animals and now that we have almost got rid of them, you want to bring them back." Surprisingly, there was also opposition from some environmental and animal-rights citizens. They objected to potential suffering that individual animals might endure, and the fact that some might die as a result of the transplant effort. They were also concerned that not enough habitat research was conducted to ensure a successful transplant. But in November, 1998, the Wildlife Commission approved the release of lynx and established a $10,000 fund to pay ranchers for any livestock loss due to lynx predation. The Mountain States Legal Foundation then filed a lawsuit on behalf of agricultural interests to stop the reintroduction of lynx. They feared that if the lynx were brought back, it would be added to the federal endangered species list and could possibly endanger their privilege of grazing livestock on public lands. A Federal District judge dismissed the lawsuit.

Vail Associates contributed $200,000, the DOW $452,000 and Great Colorado Outdoors $130,000 of lottery dollars for lynx reintroduction. In January of 1999 trappers near Lac la Hache, British Columbia, trapped lynx for Colorado. These were hauled to holding pens at Herman and Susan Dieterich's wildlife rehabilitation center at Del Norte, Colorado. They were fitted with radio transmitters and held in conditioning pens. On February 3, 1999, the first four female lynx were released on private land at Humphreys Lake about ten miles south of Wagon Wheel Gap at the edge of the Weminuche Wilderness. The females were given a head start to establish their home territory. Soon afterwards five males were released. Alaska and Yukon Territory wildlife agencies furnished another thirty lynx that were released later in the spring at Goose Creek, Red Mountain Creek, Pierce Creek on the Rio Grande side of the Continental Divide and on Fish Creek, Vallecito, Lemon, Sand Bench, First Fork of Piedra, and near Williams Creek Reservoir on the San Juan side. Five of the first lynx died of starvation soon after they were released. The balance of the new lynx were retained longer, fed a good diet, and released in the spring and had a much higher survival rate. By the fall of 1999 the lynx had scattered as far east as northeastern New Mexico, Cortez, and one traveled from near Creede nearly 400 miles to where it was shot and killed along the North Platte River near Oshkosh, Nebraska. By the fall of 1999 about twelve of a total of forty-one had died from starvation, roadkill, being shot, or unknown causes. Most of the lynx remained in the San Juans. In January of 2000 the Wildlife Commission gave approval for the release of an additional fifty more lynx in the spring of the year 2000. The risk of such ventures should be measured not only by the cost of failure, but by the value of the goal of successfully reestablishing the lynx in Colorado. Time will tell.

Beaver

Since the mid-1960s the state has allowed private beaver trappers to trap beaver on public lands in Colorado. The District Wildlife Manager determined how many beaver would be taken from each drainage in order to maintain a stable population. Trappers then applied for limited permits, and this system worked successfully to control beaver where they plugged highway culverts, canals and ditches. A District Wildlife Manager often had to trap nuisance beaver in the summer months, because the summer pelts were of no value to trappers. In the

1960s many trappers quit using leg-hold traps to catch beaver and started using the humane Connibear trap that killed a beaver instantly. After the nuisance beavers were caught, a wildlife officer bought dynamite and blasted the beaver dam so water would flow through the culvert or canal. There was no formal training required to use explosives. A wildlife officer learned to blow beaver dams from a veteran officer, preferably one who could still count to ten on his fingers. In the 1980s wildlife officers were trained to use newer and safer explosives, but soon the Division of Wildlife stopped its personnel from using explosives. County and state road crews or landowners had to take care of the beaver problems themselves.

Beaver management changed in 1996 when the voters of Colorado approved a constitutional amendment banning the use of leg-hold and instant-kill traps and for all practical purposes took away the most practical tools to manage beaver and other nuisance animals. The writers of the amendment gave little consideration to the plight of landowners who have to contend with beavers plugging ditches, and culverts, and cutting down desirable trees.

Jerry Apker, Area Wildlife Manager for the San Luis Valley, described the situation:

> We no longer harvest beaver to manage populations. When the population builds in a few more years, beaver will become more of a problem that we can't do much about. Even on our wildlife areas the beaver is taking more of our time, not to kill them, but to remove debris from diversion structures. We are modifying many of our water structures to make them more resistant to being plugged. Landowners can come to us and request a thirty day trapping permit only once a year for beaver control. They must exhaust all other attempts to control the beaver, before we can issue a permit. If a county calls us about a beaver plugging culverts, we have no recourse. No one can legally trap the beaver. Beaver can be shot, but they usually don't come out during daylight hours. Trapping permits can be issued only to control nuisance animals that are interfering with agriculture.

Managing Lands for Wildlife

So much critical wildlife habitat has been lost in Colorado that the Division of Wildlife has purchased a number of properties. These state wildlife areas are intensively managed to enhance wildlife. For

Mike Zgainer fills a fertilizer spreader on the Coller State Wildlife Area to improve range conditions for elk. (January, 1975). Photo by author

example the Coller State Wildlife Area was bought in 1964 to provide public access to the Rio Grande and for elk and deer winter range. At that time sportsmen were demanding more access across private land to the public lands so the Division of Wildlife built a public access road around private land to Haney Gulch and other public lands. Coller State Wildlife Area was a sheep ranch for many years, and the forage had not recovered. Wildlife Conservation Officer Mike Zgainer got the idea to improve that critical winter range that was also a migration corridor. Zgainer said:

> I read an abstract of a research project that George Bear, Wildlife Researcher, had done in the La Garita Mountains on bighorn sheep in the 1960s and '70s. Bear's study showed that even though he was only fertilizing bighorn range, the elk were drawn to his fertilized plots. Since the objective of the DOW at that time was to increase elk, I decided to experiment. I contacted Bob Clark, our Regional Habitat Biologist, and he got the funds to purchase the fertilizer with Pittman-Robertson federal aid money.

The Coller State Wildlife Area is an especially important corridor now , since there is extensive subdivision going on above and below that property. It was a major success to have the area go from supporting seventy-five head of elk to over 400 head which spend several weeks there each fall, before migrating across the Rio Grande to winter ranges north of South Fork.

A small herd of elk beds down for the afternoon at the Coller State Wildlife Areas. As many as 400 head of elk have been seen in the area during the migration to their winter range. Photo from Mike Zgainer Collection

To further enhance the winter range the Division of Wildlife built an irrigation system to increase forage production. A water pump lifted water from the Rio Grande onto flat areas where a substantial improvement in winter range forage was realized.

Since the success of Zgainer's fertilization experiment, two other fertilization projects were completed on Forest Service lands on San Francisco Creek south of Del Norte and Farmers Creek east of Creede. Both areas are isolated and had no vehicular access so the Division of Wildlife hired a helicopter to spread ammonium nitrate fertilizer on portions of these critical winter ranges. The effects of these projects were not as dramatic as at the Coller and have not been continued.

Fire has been used as a tool to stimulate decadent forage by recycling nutrients on many wildlife areas in Colorado. In 1987 Dave Kenvin set a prescribed fire on the Coller area. In the early 1990s Tammy (Fox) Spezze, Creede District Wildlife Manager, and the Forest Service burned a small area to improve bighorn sheep range on Long Ridge. It is difficult to measure objectively such attempts to improve range, but forage production does increase and the animals seem drawn to those areas at least for the short term.

Jerry Pacheco, District Wildlife Manager in South Fork, said:

> We found that we can fertilize about every third year and irrigate as needed. We don't burn much anymore, but we mow what the elk don't eat such as the big rabbit brush and then fertilize it and they are starting to eat it. We still have a lot of elk spend time on the Coller during their migration.

Such land management practices have improved wildlife habitat in Colorado. Although many were designed to benefit primarily deer, elk, and bighorn sheep, such range improvements also improve habitat for smaller animals and birds. Projects often involve multi-agency cooperation as well as private landowners.

Forest fires have been suppressed on most public lands, and the San Juans have been no exception. Fire is an important factor in natural ecosystems. The Rio Grande and San Juan National Forests have a policy outside of the wilderness areas that allows lightning-caused fires to burn within prescribed areas before suppression is initiated. This allows wildfire to play its role on selected sites to improve wildlife habitat and maintain the natural diversity of the forest. Man-made fires are attacked and suppressed immediately.

Antelope

Although antelope had been reintroduced to the San Luis Valley in the 1940s, some herds were not sustaining themselves. One of Punk Cochran's last projects in the South Fork District was the transplanting of seventy-five antelope which were brought down from northwestern Colorado and released northwest of Del Norte to supplement a small herd. That base herd increased to nearly 400 head and sustains limited hunting. Antelope have not been reintroduced to the Upper Rio Grande because of development, fences, and subdivisions. In the summer of 1995 two antelope showed up on Long Ridge, but it is unknown where they came from or what happened to them. Antelope were planted on the Cochetopa west of North Pass in the 1970s to supplement a native herd in that area. Severe Gunnison winters are a major limiting factor for the antelope in that area. They do very well for a few years and then a devastating winter can wipe them out.

Bighorn Sheep

In the 1970s bighorn herds were doing well around Ouray, the Almont Triangle, Trickle Mountain, the Sangre de Cristos, and in several other small herds on the west side of the San Juans. The only bighorns left on the Upper Rio Grande were on Pole Mountain and the La Garita mountains. During this time Division of Wildlife researchers had developed treatments for the lungworm parasite and Pasturella pneumonia that was responsible for the decline of the bighorns. Because the Division of Wildlife was able to treat captured bighorns for these maladies, it was successfully reintroducing bighorns into many historic ranges in Colorado.

In the mid-1970s Mike Zgainer initiated an Environmental Assessment for a bighorn transplant north of Del Norte, but it wasn't until 1982 that Dave Kenvin, South Fork District Wildlife Manager at that time, witnessed the transplant of twenty bighorns to the Natural Arch area. That transplant was successful and the population steadily increased and provided some limited hunting. But Jerry Pacheco, South Fork wildlife officer, said that the herd crashed in 1996 from an outbreak of Pasturella pneumonia. Nearly seventy percent of the bighorns died. They most likely contracted the disease from domestic sheep that grazed nearby. The herd is slowly recovering.

Creede produced a native son, Dale Hibbs who became one of the top sheep and goat biologists in the country. Hibbs grew up in Creede and Salida. The author and he were classmates at Colorado State University, and when Hibbs came back to visit his roots in Creede their conversation always turned to bighorn sheep. Dale planted the seed of desire to someday reintroduce the bighorn to the Rio Grande. Dale and his father Ashby were killed in an airplane crash while on a hunting trip in Glenallen, Alaska, in 1967. In 1977 Hibb's idea germinated when the author initiated the Environmental Assessment with the Forest Service to reestablish the bighorn at five locations in the Upper Rio Grande. Before proceeding the Forest Service required documentation that bighorn had lived at the locations where transplants were being proposed. The author interviewed many of the people who remembered the bighorn sheep at the turn of the century. Research biologists George Bear, Bill Rutherford, and Bob Schimdt evaluated these historic ranges and prioritized each site. In addition to checking out the sites, public meetings were held to get input from ranchers and others who might have had concerns about bighorn reintroductions. It took six years of paperwork and waiting for the Upper Rio Grande to become high enough on the statewide priority list. Ken Miller, the District Wildlife Manager in Ouray, trapped twenty-two

Three radio collared bighorns run to cover in this January, 1983, release of twenty-two bighorn sheep on Seepage Creek. This transplant was the first of five such reintroductions of bighorns to the Upper Rio Grande. Photo by author

bighorns at Jackass Flat above the town of Ouray. The sheep were treated for lungworm and released southwest of Creede.

As with all reintroductions, some of the ewes were fitted with radio collars, the rams ear-tagged and lambs were neck-banded. The herd scattered and weren't one bit afraid of human activity. Within a few days the author got a call from David "Jake" Powell at Spar City, "We just watched one of the bighorns walk up to one of the summer homes and kick the glass door, shattering it." A few days later Darcy Brown called from Humphreys Lodge, "Ruth and I just had a bighorn sheep walk across our patio so close to us that we could read his ear tag-number "six." This herd was supplemented the following year with a transplant from the Almont Triangle north of Gunnison. These bighorns were released at Lower Wrights' cabins on Colorado Highway 149. Soon the two transplants found each other and ranged from Long Ridge to Bristol Head Mountain. These bighorns have not reached population levels that will sustain any hunting but have provided wildlife watchers and photographers many hours of enjoyment.

By 1987 the Seepage/Long Ridge herd had not increased. Lamb survival was poor with lungworm being suspected. Several bighorns had been seen coughing, and pneumonia could easily spread through the herd. A drop-net trap was baited near Highway 149 at Lower Wrights' Ranch. When most of the bighorns were coming into the trap, it was dropped and twenty-two head were treated with antibiotics and worm medicine and then released. This was done to improve the health of the herd and boost lamb production. Many Creede volunteers assisted the biologists and a veterinarian to restrain the animals for treatment.

In 1985 twenty bighorns were transplanted from Cottonwood Creek west of Buena Vista, Colorado to the confluence of Blue Creek and the Rio Grande. A second plant was made in 1988. These transplants were the most successful in the Upper Rio Grande, and a very limited hunting season was allowed in 1990 with one permit being issued. The herd built up to about 150 head, but in 1996 a Pasturella pneumonia infected the herd. Tammy (Fox) Spezze, District Wildlife Manager, said:

> It was really tragic. When the bighorns started dying I hiked all over the area from Blue Creek to La Garita Ranch and I found about 100 carcasses. I gathered various organs and had them analyzed by Mike Miller, a veterinarian at Colorado State University. He found some lungworm infestation, but the cause of death was Pasturella pneumonia. It was most likely transmitted from domestic sheep. Only about forty bighorns survived. I did put out some medicated feed blocks and I think it helped. There was also a die-off of bighorn in the La Garita Wilderness about that same time. I had seen bighorns and domestic sheep together in that high country just before that die-off.

The Blue Creek herd is slowly recovering.

The bighorn herd on Pole Mountain had up to one hundred head prior to the 1950s. By the late 1950s the population crashed, and since then a small herd of fourteen to twenty head is all that the mountain sustains. Before doing anything to increase the herd it was necessary to determine if the lungworm parasite was a limiting factor for these bighorns. Jim Olterman, Wildlife Biologist; George Wiggins, Forest Service Biologist; citizens Rod Wintz and Rex Sheppard, and the author climbed Pole Mountain to gather samples of bighorn droppings. Bill Adrian, Wildlife Researcher of the wildlife research center in Fort Collins, examined the samples and determined that lungworm was present in sufficient density that it could be a limiting factor.

If only a site could have been found where the animals could be baited, trapped, and treated then there would be hope that the herd could be increased. Ralph Conkey and the author made several snowmobile trips to the base of Pole Mountain to look for wintering bighorns. In addition helicopter and airplane flights were made to determine the extent of the winter range these sheep utilized. Amazingly the Pole Mountain bighorns stay year-round on Pole Mountain at elevations to over 12,000 feet and often in very deep snow. The Division of Wildlife hired a helicopter to place salt blocks in strategic locations on Pole Mountain to see if bighorns could be baited to an accessible location. One site was really spooky. A very skillful pilot maneuvered a Hughes 500 chopper to the edge of a cliff, putting one skid on the cliff while keeping the chopper flying as the author got out, opened the rear door, pulled a fifty pound salt block out and left it on the edge of the cliff. Other salt blocks were placed at better landing sites. The bighorns came to the salt blocks, as did elk and deer, but no place could be found where they could be baited and treated.

The last attempt to determine the extent of Pole Mountain bighorn range and to find a winter location where the Pole Mountain bighorns could be treated was in 1987. Three bighorns were trapped from Trickle Mountain west of Saguache, radio collared, and taken to Rio Grande Reservoir, where they were flown by chopper to the "castle" of Pole Mountain. Don Masden, Wildlife Biologist, remembered:

> The chopper left me on the edge of a cliff on Pole mountain and returned to the dam at Rio Grande Reservoir. He was gone a long time, long enough that I got to thinking I was a long way from civilization and if he didn't come back I would have a long walk in deep snow to get off that mountain. The chopper made three trips to bring one ram and two ewes, each in a separate box, suspended under the chopper. When I had all three boxes I opened them and the sheep ran up the mountain. That southern exposure was windswept and the snow wasn't very deep, but it was really deep in the timber. Jim Olterman, our regional biologist and airplane pilot, and I tracked the three

sheep with radio telemetry. They didn't go far. The ram and one ewe died of unknown causes the following summer in West Lost Trail Creek. The third ewe was last seen on the southeast side of Pole Mountain in August of 1988. The radio quit soon after that and it wasn't heard from again.

Pilot Richard Dick flies a bighorn sheep past Rio Grande Reservoir on its way to be released on Pole Mountain. Photo by Don Masden

Later Masden released two other bighorns in the hope that they would join up with the Pole Mountain herd from the north. They were released at the mouth of Mill Creek southwest of Lake City. Both of these attempts failed because the released animals all died before any information could be gathered. Unless the bighorns move or other information changes, this herd is so isolated in rugged terrain that it is impractical to treat it for lungworm. The herd may continue to sustain itself, or it could cease to exist. Domestic sheep have not been grazed on Pole Mountain since the 1920s, yet the bighorn population remains suppressed. The fact that the mountain sustains a larger-than-ever elk population may play the most important role as the current limiting factor.

The last bighorn sheep transplant in the Upper Rio Grande was made in 1989 at the top of River Hill at the head of the Box Canyon, two miles below Rio Grande Reservoir. That herd scattered; however, bighorns are now seen in that general area.

On the east side of the San Luis Valley is the 600,000 acre Forbes-Trinchera Ranch. Conrad Albert, District Wildlife Manager, said that in the mid 1980s ranch manager Errol Ryland negotiated with the Province of British Columbia for bighorn sheep in exchange for making a donation to the province. Thirty-four bighorn were shipped to Colorado. They were inspected for diseases and parasites by Mike Miller, Colorado State University veterinarian, and released on the ranch. As soon as they touched Colorado soil they became the property of the State of Colorado. An agreement through Colorado's Ranching for Wildlife Program allowed the state and Trinchera Ranch to work together for a hunting agreement. Ninety-percent of the licenses go to the ranch, and ten percent are reserved for the general public. The ranch has removed domestic sheep from the habitat and conducted extensive burns and other habitat manipulation to benefit the bighorn. By 1999 the herd has increased to over 200 head, and in 1998 the new state record for bighorn was taken by a private hunter on the ranch.

Bighorn sheep have lived in southwestern Colorado for thousands of years, but the activities of man changed their habitat and stressed them in other ways as well. Pasturella pneumonia, passed from domestic sheep, has been suspected for causing the decimation of some herds and competition for forage with domestic sheep and elk have limited their nutritional base. Bighorn are also susceptible to their own population density and if allowed to overpopulate, an entire population can die off. Hunting and live trapping have been the two most successful management tools used to perpetuate sustainable herds. Yet Colorado's State mammal continues to sustain itself. In 1999 the Division estimated that there were more than 1,700 bighorn sheep in southwestern Colorado accounting for nearly a fourth of Colorado's bighorn population.

Moose

In 1976 the Wildlife Commission requested an evaluation of potential moose introduction sites in Colorado. Because of its extensive willow-bottom areas, the most preferred site in Colorado was in the Taylor Park area in the Gunnison Basin. Local ranchers, however, were opposed to stocking moose in that drainage, and North Park became the first transplant site in Colorado in 1978. An occasional Shiras moose had wandered from Wyoming into North Park, but there was never a viable population. The Shiras moose, which is smaller than the Alaskan moose, lives in a variety of habitats, although it prefers willow bottom lands. Even though moose were never native to the Upper Rio

At two o'clock on a cold morning in February, 1992, a young moose wakes up to find Jay Sarason opening the trailer door to its new home in the Upper Rio Grande. Photo by author

Grande, there was a habitat niche for them that they could have inhabited naturally, except for the invasion of the white man.

In 1990 the Rio Grande National Forest Plan directed the Forest Service to determine if there was moose habitat on the Rio Grande National Forest. It is uncertain how the moose ever became a part of the forest plan. But from this the Forest Service and the Division of Wildlife developed a plan to introduce moose into the Upper Rio Grande. That plan included gathering public comments to see if there was public support. Ranchers in North Park had not experienced any damage from moose, so there was no opposition from ranchers in the Upper Rio Grande. Between 1991 and 1993 the Division of Wildlife, Forest Service, and a host of volunteers hauled a total of ninety-two moose from north central Colorado, Utah, and Wyoming and released them at twelve locations in the Upper Rio Grande. Some moose were captured by shooting them with tranquilizer guns, but most were captured by shooting them with a net-gun from a helicopter.

Jay Sarason said:

> The first part of the transplant came from North Park near Walden. We had so many volunteers that I'd hate to list them, because

Two bull moose square off in their new home in the San Juan Mountains. Photo by author

someone might be left out. Volunteers brought trucks and horse trailers and we drove up to Walden. A blizzard hit and we were stuck there until the storm broke. Then came the slow process of hunting and tranquilizing the moose. Herman and Susan Dieterich, wildlife rehabilitators in Del Norte, were of immense help in providing us their expertise to manage the captured moose. After we got loaded up, we drove all night to get to Creede. Dave Kenvin and I had a big bull in the back of our rig. We were so tired when we got to the Brown Lake turnoff on Highway 149, we pulled over and planned what we would do. We left both doors of the pickup open so that we could escape to the truck, because we knew he was going to be on the fight. So we opened the trailer door very quickly. Kenvin ran for the passenger side of the pickup, but when I ran for the driver's side I knew the moose was going to get me first, so I jumped up on top of the pickup. It was like a scene out of 'Jurassic Park'. That big bull stuck his head into the driver's side and was eyeball-to-eyeball with Kenvin. The bull quickly had enough of Kenvin and just backed off and ran down into the willows. That was when I started breathing again.

> We soon tracked the moose by radio telemetry from both the airplane and with hand-held receivers. It is amazing how a big moose could just become a ghost in the willows.

Most of the moose stayed in the Upper Rio Grande, but a few traveled as far as 150 miles to the west and to the north. A number went over the Continental Divide and established themselves between Spring Creek Pass and Lake City. Although there have been a few accidental shootings by hunters and some lost to unknown causes, the moose population had grown to over 375 head by the summer of 1999 and had dispersed into several other drainages in southwestern Colorado. Funding for the project came from the Division of Wildlife, Safari Club, and the Farley Foundation. The first limited moose hunting season was in the fall of 1999 and two bull moose were killed by hunters.

Effects of Logging, Grazing, and Roads

Logging and its attendant road systems have greatly affected big game for the past one hundred years. Roads increased vehicular access into previously remote and wild mountains. Elk were pushed into even more remote areas. It wasn't until the 1980s that wildlife became a priority for the Forest Service. Before that time the Forest Service designed timber sales to maximize the number of board feet of timber that could be cut. The forests were being managed like tree farms for maximum wood production and not as an ecosystem. Even with the National Environmental Policy Act process of obtaining input from local communities and the Division of Wildlife, the preferred alternatives that would benefit wildlife or at least minimize the impacts of such activities upon wildlife were seldom chosen.

The Division of Wildlife has no veto power over Forest Service operations, or any other entity for that matter, but the Division of Wildlife asked for reconsideration. The Forest Service did try small patch cuts that fitted into the landscape and created openings in the forest. These openings were also to serve as snow basins to increase runoff from the snowmelt. These smaller clear cuts were landscaped so that from a distance they were barely noticeable and less objectionable to the public. Aspen and other forage invaded the new openings, and elk were initially drawn to them like steel to a magnet. Timber cutting operations were soon designed to be wildlife habitat improvement projects. Some considered this another excuse for cutting more trees and building more roads into roadless areas, as there was a lot of habitat that didn't need improvement.

The public became very alarmed at what logging was doing to the forests throughout western Colorado. Citizen groups began to form and oppose the Forest Service timber management practices. Some groups became radical, but in the Upper Rio Grande local people started the "Creede Timber Watch." This was a diverse group of citizens whose purpose was not to stop logging, but to demand that logging operations be more sensitive to the ecosystem. Jay Sarason, Creede District Wildlife Manager, said:

> We met with the Forest Service and pointed out that they were not following their own timber management plans and were not policing the timber harvest. Some of their own people felt the same way. We changed how they logged timber to leave trees for thermal cover, stream protection, to not log steep slopes and to maintain parks. I don't believe this group was really extreme, but we just wanted more consideration for wildlife. My role was more advisory as at that time DOW personnel were supposed to be "value neutral." In my opinion we quit being an advocate for wildlife and were just advisory and there is a big difference

Jim Webb, Forest Supervisor, skillfully led the newly consolidated San Juan/Rio Grande National Forest to listen to the various interest groups and consider how decisions affected people and the forest and to educate the public about land management issues. His desk was often a big rock out in the forest where he listened to his professional staff, forest users, environmentalists, and the forest itself. Webb said, "Out there in the woods people can solve problems together much better than in a stuffy office with no windows." He had an ability to bring people together who didn't appreciate each other's position and develop consensus for more acceptable courses of action. As a result of his leadership the Forest Service changed timber harvest practices for Engleman spruce from clear cutting and shelterwood systems to mostly a "selective group cut" wherein small groups of trees were selected and cut in such a way as to maintain a cover of large trees as

Jim Webb, Rio Grande Forest Supervisor, at home in his "office." Photo from Jim Webb Collection

well as a diversity of younger age classes. This system was more difficult to plan and manage, but it was a compromise between the demands of environmentalists and the needs of the timber industry. Webb initiated a new era of public involvement that some extremists on either side of such issues may never appreciate, but the Forest Service honestly listened and considered their viewpoints in the decision making process.

Throughout Colorado some organizations take their public involvement beyond their vocal consensus with cash and labor. For example in 1994 the Rocky Mountain Elk Foundation donated $3,000 to the Forest Service to buy grass seed for a revegetation project. Under the supervision of the Forest Service the Division of Wildlife furnished equipment and personnel to disk and harrow several miles of old logging roads on the northern slopes of Baldy Mountain and planted high quality grasses. The roads were closed to vehicular traffic and designated as wildlife habitat restoration areas. Such cooperative projects improve habitat for wildlife and stabilize soils following a logging operation. Elk immediately discovered these successful plantings and made good use of them. This is an example of the good that small projects can do. There have also been larger projects and purchases of land by the Nature Conservancy and Rocky Mountain Elk Foundation of critical habitats in the Gunnison, San Miguel, and San Juan Basins. Throughout western Colorado there are local land trust organizations that are purchasing open space. Together, they are saving what they can of critical wildlife habitat.

The Forest Service changed grazing practices on the San Juan/Rio Grande National Forest between 1980 and 1999. The sheep industry had been in an economic decline for the past twenty years. As a result, all of the domestic sheep allotments in the Weminuche Wilderness have been closed to domestic sheep grazing. Some allotments outside the wilderness have been abandoned, and some of the remaining woolgrowers have been moved to different allotments. The Forest Service has, where possible, removed domestic sheep from proximity to existing bighorn sheep herds. This should give a major boost to the health and recovery of the bighorn in the San Juans and other areas where there has been conflict between bighorn and domestic sheep. Cattle numbers have gradually been reduced on many allotments and some allotments have been closed altogether. There is still a need to evaluate the grazing of some critical winter ranges, because they are generally not up to a desirable condition. Light to moderate grazing can be beneficial to overall range health, but heavy grazing has been devastating to native vegetation, soils, and riparian areas.

Grazing practices are also being modified in the Gunnison Basin to protect the Gunnison sage grouse, a species that could soon be listed as endangered. Ranchers, the Division of Wildlife, the Forest Service and the Bureau of Land Management are working together to protect the Gunnison sage grouse breeding grounds. In addition there have been other adjustments to grazing practices to enhance winter range for deer and elk.

The proliferation of roads into previously roadless areas throughout western Colorado has caused an increase in unsportsmanlike road hunting. Most hunters didn't want more vehicular access, but if someone builds a road, why would anyone want to walk? Some hunters gradually started teaching young hunters how to operate a 4-wheel drive or an all-terrain vehicle rather than the skills of the woodsman-hunter. The Forest Service and Bureau of Land Management started to

Four-wheel drive vehicles tore up several areas that led to the Forest Service restricting vehicles on national forest lands in Colorado. Photo by author

place restrictions on off-highway vehicle use by first closing extraneous roads and those that were causing soil erosion. Gradually many of the old timber sale roads were physically closed or gated, and elk once again returned to escape cover that was less accessible to the public. New timber sale proposals still had road construction, but they were closed to the public. Presently, vehicular access is restricted in most areas to existing roads and trails, and cross country travel is prohibited. Vehicle use has had a detrimental effect on wildlife and yet it is essential so people will have access to different areas and for the management of forest resources.

Travelwise, the San Juan mountains have an excellent diversity of opportunity. All of the wilderness areas allow only foot and horseback travel. But in other areas there are many miles of trails accessible to trail bikes and all-terrain vehicles that lead to secluded hunting and fishing areas. There are four-wheel-drive roads that offer challenging trips and spectacular scenery, hunting, and fishing. Yet, some of the most popular fishing is accessible by passenger car or motorhome. Regulating human activities to minimize the impacts on wildlife and the land is a controversial issue. Vehicles have should have their place, but not necessarily be in every place. As the people's representative of the largest landowner in southwestern Colorado, the Forest Service has listened and responded to the public, but also has been sensitive to balancing environmental needs with public demands.

Weather

The San Juans, like all of Colorado, are susceptible to climatic extremes. The 1975 winter was very dry. There was practically no snowfall. Fishermen were able to drive their vehicles out on the ice at all the lakes and reservoirs that one can normally drive to only in the summer. In the spring of 1975 there was no runoff and irrigation water was limited. By late summer the Rio Grande was flowing about sixty-two cubic feet per second at the Del Norte gaging station when it would normally flow at ten times that volume. The monsoon rains of late summer began and the river rebounded enough to keep the fish alive, except in some of the highest streams.

Drought conditions were broken in the 1978-79 winter. It started snowing on Halloween night and it snowed all winter. Temperatures reached minus fifty-four degrees in Creede, and the snow was deeper than the fence posts all winter. The weight of snow caved in roofs. The state and county road crews plowed snow daily. People had difficulty getting around and the utility bills were higher than normal, but everyone pulled together and survived the winter.

Wild animals of the San Juans were designed to survive snow and cold, but once again winter was about to deal a death sentence by starvation. Once again the two edged-sword of winter feeding would be drawn in the Gunnison and Upper Rio Grande. Elk were in trouble by Thanksgiving. Herds soon ate the forage where they were and the snow was too deep for them to move. The early snows melted down to about a foot of ice and hard crusted snow, but the snow continued to pile up. Deer and elk were facing the same conditions in the Gunnison. It soon became obvious that there would be a large die off if the animals were not fed. The decision to start feeding in the Upper Rio Grande came early in December and continued until late April. Over 125 volunteers fed about 1,200 elk every day on fifteen feeding grounds. The Division of Wildlife bought high quality hay from San Luis Valley ranchers as well as high protein pellets that had

In thirty below zero temperatures, in January of 1979 volunteers scatter alfalfa and pellets for elk on the La Garita Ranch. Photo by author

been improved since the 1968 feeding operation in Creede. Men unloaded semi-trailer loads of hay, sometimes after dark, and often in minus thirty degree temperatures. Others hauled hay to feeding areas once a week and still others scattered the hay to waiting elk every day. There was still no ration available that deer could digest so the mortality of deer was extremely high. Snowmobilers found one herd of elk on the trail into Wheeler Monument. The Division of Wildlife dropped hay from a chopper and later took hay into Trujillo Creek by snowcat. Snowmobilers helped scatter a trail of hay down to Blue Park where the elk worked their way down to join elk being fed on the La Garita Ranch. Paul Hosselkus, the Mineral County Road boss, saw to it that the feed grounds at different ranches were plowed out so that the hay could be scattered more efficiently. Volunteers fed out over 300 tons of hay and over ten tons of pellets before south-facing slopes cleared enough in late April for the elk to go back to their high country.

The Gunnison Basin is one of the coldest valleys in the lower forty-eight states. Cold air settles in the valleys all winter, and this severe cold saps the fat reserves of all wildlife. Only the tough survive. In the 1978-1979 winter a similar feeding operation was conducted to

Creede High School students listen to George Wiggins, Forest Service biologist, explain the plan to search for and record dead elk and deer in an area of Farmers Creek. Upper Rio Grande outfitters furnished horses to take students to the transect site and cooked up a lunch afterwards. (May, 1979) Photo by author

save deer and elk in the Gunnison Basin. Elk were successfully fed in those areas where they could be reached, but mortality was higher for deer as a digestible ration had not yet been developed for them.

The Division of Wildlife, volunteers, and high school students walked and rode around the feed grounds and nearby areas looking for carcasses and found that about fourteen percent of the adult elk had died as well as nearly half of the calves. Bill Adrian, Wildlife Researcher, and Beth Williams, veterinarian from Colorado State University, necropsied several adult and calf elk that died on the feed grounds. They determined that the dead calves on the feed grounds died from hypothermia. Williams explained, "These calves are so small that when it gets this cold [minus thirty to minus fifty degrees], they cannot burn enough fuel in their little 'furnaces' no matter what you feed them, to generate enough body heat to survive."

If the Division of Wildlife and the citizenry had not fed elk that winter, most of the elk would have died in the Upper Rio Grande. The elk herd recovered within a couple years, but the deer herd took longer.

When the 1984 winter came the Upper Rio Grande conditions were not as severe as in Gunnison and feeding was not necessary. But the Gunnison Basin and several areas in northern Colorado conducted one of the most extensive winter feeding operations in the state's history. Alfalfa was dropped from helicopters to isolated herds. Hundreds of volunteers assisted Division of Wildlife personnel by hauling and scattering alfalfa and hay. They also successfully fed thousands of deer with newly developed pellets that deer could digest. The operation cost nearly six million dollars; saved nearly ninety-five percent of those fed, but only reached about five percent of western Colorado's deer. After two years the deer herds had nearly recovered in most areas.

In 1996 the Gunnison area experienced another severe winter and over 2,500 deer, 5,000 elk and several hundred antelope were fed at twenty-five locations. The feed alone cost over $250,000. Major General Westerdahl of the Colorado National Guard loaned six articulated snowcats to the Division of Wildlife to distribute pellets and alfalfa. Nearly 300 volunteers donated time and work to feed the animals. Jim Young, Area Wildlife Manager for the Gunnison Basin said, "When conditions got tough, the volunteers really pitched in. We couldn't have done it without them." There was very low mortality among the animals that were fed. However there is one cost to winter feeding that is becoming more and more apparent. It costs the animals their wildness. They become dependent on mankind for their food, because humans are destroying their habitat.

Elk Management

While the elk herd was recovering from the 1978-79 winter, the hunting pressure was increasing dramatically, and the harvest of bull elk left very few mature bulls in the base herd. The winter helicopter elk counts verified that there were very few mature bulls after the hunting season. However, the number of cow elk and calves was increasing, even with the low bull ratio. When game counts are made the biologists classify only a sample of the population. Every attempt is made to count the same areas in the same way under similar conditions, usually deep snow. In this way it is not necessary to fly every acre and some animals are missed by even the best observer. Computers provide more accurate and faster calculations of population dynamics, but the data that goes into the computerized models must be as accurate as possible.

Statistics are boring, but counting thousands of elk in a few days is pure pleasure punctuated by moments of hair-raising terror. In February of 1982 Larry Nelson, a pilot for Roberts Aircraft, was flying local wildlife officers for the annual winter elk trend and sex ratio counts. Nelson was flying a powerful Hughes 500 helicopter. The day after he had completed the counts on the Rio Grande side of the San Juans he was flying Pagosa Springs District Wildlife Manager Glen Eyre and his wife Linda who was an observer. While they were looking for elk the engine governor failed and they made a forced landing in the high country east of Pagosa Springs. No one was injured. They radioed for help and another helicopter in the area responded and rescued them.

About three days later Nelson was flying a replacement helicopter with area biologist Don Masden and Herb Browning. They had counted about 500 elk during the morning flight and were flying the afternoon count south of Williams Creek Reservoir. Suddenly the coupler in the tail rotor drive shaft snapped and the chopper immediately started rotating out of control. The chopper dropped straight down and crashed into a tiny clearing in about five feet of snow. The impact was so hard that the rotating blades hit in the snow and snapped off as did the tail boom. The cabin of the Hughes 500 absorbed the crash quite well.

Although they survived the crash Nelson and Masden were in excruciating pain from compression fractures of their spines. Browning wasn't seriously injured. The emergency locator transmitter began broadcasting and a passing airliner over heard the signal and relayed to Denver Stapleton Airport that they were picking up an emergency signal. Browning heard the pilot and radioed to him that they had crashed at the junction of Sand Creek and the Piedra River and that two were hurt and one was okay. Browning closed his transmission with, "Tell them to bring burgers and fries when they come to get us." People back in Pagosa Springs would know then that even though there were injuries they were probably okay. The pilot relayed Browning's message verbatim to Denver, but it would be daylight the next day before the rescue operation would be conducted.

Darkness quickly came. The temperature dropped to below zero. The cold temperature caused the battery to go dead, so there would be no more two-way radio transmissions. With only one survival kit aboard they put the pilot in a sleeping bag in the front seat. Masden laid down in the back seat and covered himself as best he could with coats. They knew they had back injuries and realized that they shouldn't be moved. Browning spent the night outside the chopper keeping a fire going. The only problem was that the snow was so deep that the fire kept melting itself down toward the ground and gathering firewood in the dark in the deep snow was very tiring. He cut some spruce boughs to throw on the fire the next morning to make smoke so their rescuers could find them. The night passed slowly.

When dawn broke a thick fog blanketed the area. The three men could hear a small airplane flying around above the fog, but observers couldn't see them. Some rescuers led by former Wildlife Conservation Officer Judd Cooney, who still lived in Pagosa Springs, had started a ground rescue attempt and had hiked all through the night down the rugged Piedra River canyon. They got to the scene just as the aerial rescue started. When the fog lifted a "Jolly Green Giant" helicopter from Kirkland Air Force Base in Albuquerque, New Mexico flew by, but they didn't see the flares that Browning had shot into the air. The smoke from Browning's fire stayed low to the ground and the rescue crew couldn't pinpoint their location. Finally, with his last flare Browning waited for the chopper to circle back and when it came into view he fired almost hitting the windshield of the rescue helicopter. They couldn't miss seeing it that time.

Since there was no place to land the chopper hovered while two paramedics rappelled to the ground to give aid and prepare the three men for evacuation. Browning was hoisted 100 feet up to the hovering chopper and brought into safety. He sat down and they put ear muffs on him to protect his hearing from the awesome noise of the straining chopper. While he waited he watched as the crewman hoisted Masden who was secure in a Stokes Litter Basket.

Masden said:

> They were hoisting me up. I could see they were just getting ready to swing me into the chopper when I suddenly dropped. My fall was controlled except for the last fifteen feet, because they cut the cable

and I went into a free fall down through the trees and landed on my back and then the litter rolled over so I was face down in the snow. The litter protected me pretty well, but I couldn't move. I was okay and the paramedics quickly got to me and rolled me onto my back.

Browning remembered with emotion:

The hydraulic system suddenly failed. I could smell hydraulic fluid. Then the bracket that held the cable out away from the chopper broke. I saw Don fall. In my mind I just knew that Don had been killed. I was beside myself with grief. We had to get that crippled chopper out of there or all of us could have been killed. We flew away leaving Don and Larry. We barely made it to the Pagosa Springs airport and the whole town was there. When we landed I asked the crewman if Don had been killed and he said, 'No-he's okay.

He wept with joy. (Herb Browning is the kind of man that handles the most life threatening situations with extreme calmness. He has survived a near fatal car accident, his pack train being struck by lightning, and has risked his own life to save others. Browning handles all this humbly by saying, "Some say God has really looked out for me. Others suggest He may be ticked off at me." Surely it is the former. Over the years many a wildlife officer has contributed to this legacy of courage and service.)

Later that day the Air Force sent another rescue helicopter and that crew completed the rescue without any problems. Masden and Nelson were taken to the Pagosa Springs Medical Center where they were examined and released. Both men went through months of pain and eventually recovered. There have been other biologists, wildlife officers, and pilots who have crashed and some have been hurt while doing census work in Colorado. Yet for the thousands of hours that have been flown the safety record is good. The helicopter is the best tool ever used to census big game animals. The procedures and techniques of counting them have been modified over the years and have provided the best wildlife inventory we've ever had. Aldo Leopold would be proud.

In the early 1980s Colorado elk hunters were complaining about the deterioration of the quality of hunting. The crowding of hunters

A herd of elk spread out on Long Ridge as a leased Division of Wildlife chopper flies overhead. Each winter a biologist and district officer counted and classified the elk to determine herd size, population trend, and ratios of cows, calves, and bulls. (circa 1972) Photo by author

increased competition for a successful hunt and detracted from their hunting experience. They also wanted to have elk herds managed so that there would be more mature bulls. The Wildlife Commission responded by establishing twenty-two quality elk units where the number of licenses were restricted and provided for uncrowded hunting and the opportunity to see more mature bulls. Game Management Unit 61 on the west side of the Uncompahgre Plateau had been designated as the only quality elk unit in southwestern Colorado. The last quality elk hunting unit in Colorado was established in Game Management Unit 76 on the Upper Rio Grande. Prior to establishing the unit, most hunters and the general public gave their support, drastically reducing the number of hunters. Outside that area the Wildlife Commission established antler restrictions to increase the number of bulls, but not necessarily large mature bulls. There was a cost to limiting hunters. The state lost license revenue, but those who made the biggest economic sacrifice were Creede and South Fork businesses including the resort operators and outfitters. Hunting pressure was reduced by nearly seventy-five percent with limited licenses on bull and cow elk in all seasons, including those for archery and muzzleloader hunters.

In 1984 the Upper Rio Grande had its first restricted season. It took several years before the regulation began to increase the number of bull elk . Since then the Upper Rio Grande has become one of the premier elk hunting areas in the United States. The sacrifice that local people made was the loss of the privilege to hunt in their own back yard every year. Instead they had to wait their turn in the drawings. The merchants, outfitters, and resorts that cater to hunters in small communities in or near the restricted units made the greatest economic sacrifice to provide high quality hunting opportunity. Presently it takes about four years to accumulate enough preference points to successfully draw a bull elk license, but the wait can be worth it. The increased demand for such high quality hunting could, however, lengthen the number of years between hunts in that area. Those who are fortunate enough to draw a license can hunt without interference from other hunters. Solitude is a very important element of hunting. The chance of bagging or even seeing a mature bull elk is a dream for most hunters. The majestic bull is also a target for hunters who stalk with a camcorder or camera.

In the early 1990s the Upper Rio Grande elk population did not seem to be responding in the way it was intended. The counts and harvest figures didn't seem to jibe. Jay Sarason organized some volunteers to make ground counts to check on the accuracy of chopper counts. For two years volunteers used spotting scopes to determine bull and calf/cow ratios. Each technique located elk that were missed by the other, so the combination of the two counts increased accuracy. This is still done periodically to check on the accuracy of aerial counts.

Hunting has been a way of life in the United States. Many of the residents of southwestern Colorado were drawn here by the exceptional hunting and fishing opportunities. Hunting has also been controversial and always will be. One of the major concerns about hunting is that some consider it a dangerous sport. Back in the 1940s the Game and Fish Department had safety messages printed in its annual hunting brochure. Every year there was a greater emphasis on hunter safety, but more and more hunters were being killed. In 1958

The Upper Rio Grande quality elk Unit 76 once again offers hunters and camera buffs alike the opportunity to "shoot" a magnificent trophy bull elk. (circa 1998). Photo by Carol Hinshaw

the Game and Fish Department adopted a hunter education program designed by the National Rifle Association. The department trained volunteers to teach the course. The course itself was voluntary until 1969 when the legislature made hunter education training mandatory for any person born after January 1, 1949.

In towns throughout Colorado Wildlife Conservation Officers and volunteer instructors started hunter education instruction. For example, in Creede young people initially received hunter education in their physical education class starting in 1966. Kids brought their unloaded rifles to class and learned safe handling and hunter responsibility. There was never an accident or incident with firearms. When Kelly Mortenson, the Rural Electric Association representative, was transferred to Creede he brought with him the credentials of a Master Hunter Education Instructor and his passion to teach young people safe gun handling, ethics, hunting laws, and survival. Mortenson taught hundreds of Creede young people. Alvin and Barbara Colville taught hunter education skills in the Del Norte and South Fork areas until Barbara was killed in an automobile accident. As a result of these

Richard Kolisch and Bill Moore pay close attention to the author's instruction in their hunter education class. The class safely fired live ammunition into a backstop built on the elementary school playground. (circa 1967). Photo by author

as well as the efforts of hundreds of other volunteer instructors, hunting has become one of the safest outdoor sports and Colorado has become one of the safest states in which to hunt.

The Brunot Treaty

The Utes had hunted longer than anyone in western Colorado, and they laid claim to the fact that they still had the rights to continue hunting, as was promised in the Brunot Treaty of 1874. In 1972 the Southern Ute and Ute Mountain tribes announced their intent to file a law-suit to reclaim the hunting rights granted to them by the Brunot Treaty. There were fears that having unrestricted hunting would allow the Utes to wipe out the elk and deer in the San Juans. That fear was unfounded. Rather than enter into an expensive and lengthy court battle, Harry Woodward, Director of the Game, Fish, and Parks Department, asked the Game, Fish and Parks Commission and Department to negotiate a settlement that was acceptable to the Southern Ute and Ute Mountain Tribes whilst also protecting wildlife. All parties agreed that the wildlife resource was the most important issue. The cooperative agreement allowed tribal members to hunt in the Brunot Treaty Area, which includes the San Juans, with a tribal permit but required no Colorado license. Utes are allowed to hunt during state-established seasons, and they must obey all other laws and regulations. They may hunt only for religious, subsistence, and ceremonial purposes. Utes are allowed to hunt on their reservations without state licenses, and each tribe establishes its own regulations that are similar to state regulations.

Human-Wildlife Conflicts

The impact of human activity on wildlife in the 1980s and 1990s has been dramatic. Wildlife conflicts have increased as the human population has grown. Urban people move to the mountains to get close to nature, but the complaints start coming in as soon as they name their cabin: "I want you to come up to my place and kill these pesky song-birds that wake me up every morning". . . "We have the cutest little bear coming to our cabin and we feed it dog food every day." . . . only to receive another call in a few weeks: "Come up here and get rid of YOUR blankety-blank bear. He ripped our screen door trying to get in the cabin!" The calls coming in could fill a book.

The bear-human conflict in mountain subdivisions increased dramatically in the mid-1990s. Tammy (Fox) Spezze said:

> People were leaving their garbage out and this attracted the bears. Because they associate humans with food they were losing their natural fear of man. The Division set up a "two strikes and you're out" policy. If a bear came into a campground or residence for the second time, we caught it and killed it. To save more bears from this fate I met with resort operators and home owners to teach them what they needed to do to prevent conflicts with bears: mainly, manage their trash better. People responded very positively and the complaints decreased and we saved a lot of bears.

Because the number of reported conflicts with bears was increasing, the San Juan and Rio Grande National Forests and the Division of Wildlife started a public education program to teach citizens and visitors how to live in wild country with wildlife. The Forest Service

Jerry Pacheco and Brent Woodward attend to a tranquilized black bear that was feasting in an unsecured garbage dumpster. After being ear tagged the bear was relocated to a remote area. If it shows up again in someone's garbage the bear will be destroyed. Photo by Jerry Pacheco

also required back country visitors to use bear-resistant food containers. Statewide, the invasion of people into lion and bear habitat was cause for great concern. Conflicts with these large predators as well as other wildlife species had also increased. The best way to protect the lion and bear has been to educate people to coexist with them.

The majority of wildlife species in southwestern Colorado have recovered from the brink of extinction. If our predecessors had not had the vision to bring them back, we would not enjoy the sight and sound of bighorn rams butting heads or the majestic bull elk bugle echoing across a canyon. The question now is whether or not society will have a vision and make decisions to perpetuate our wildlife heritage.

Chapter 11

WILDLIFE OFFICERS AND OTHER STEWARDS

This book was written in part to pay tribute to the hundreds of people who have been stewards of our wildlife heritage. The bulk of the examples of that stewardship have come from the Upper Rio Grande where the author was a Wildlife Conservation Officer and District Wildlife Manager for more than twenty-one years. This last chapter is about some memories, feelings, and impressions of what it was and is like to be a wildlife officer in the Upper Rio Grande and the influences people and the country had on these public servants. Undoubtedly, these perspectives are shared by most wildlife officers in small communities throughout western Colorado. If you don't know any of these people, you may become acquainted with a wildlife officer someday who would say, "Yes, that's the way it is."

Although game wardens have usually been respected, they were often not appreciated. People today have had more contact with the Wildlife Conservation Officers and District Wildlife Managers than with the early day game wardens. Regardless, most of their work has been behind the scenes and gone unnoticed. Many of our state wildlife areas exist as they are today as a result of wildlife officers and biologists having the vision to improve conditions for wildlife. They have always been fiercely independent, highly motivated, enthusiastic, and hard working individuals just like their predecessors.

Prior to 1964 a wildlife officer received little formal training but was assigned to a district. The training officers were one's neighboring officers who taught the rookie the ropes. After 1964 the Game, Fish and Parks Department hired a training officer and began intensive training for new officers as well as annual in-service training to update veteran officers. Supervisors also made assignments to broaden the knowledge of new officers. For example Ed Bechaver's experiences were typical when he transferred from the Trinidad district to Creede in the late summer of 1965. Bechaver said:

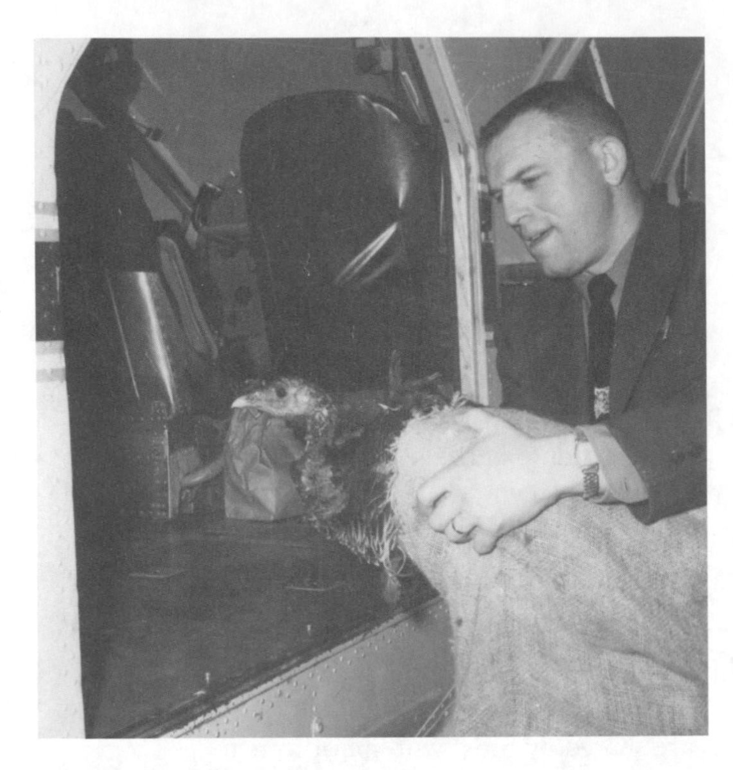

Ed Bechaver, in about 1964, assisting with a turkey transplant in the Trinidad area. He served in the Upper Rio Grande from 1965 to 1966. Photo from Ed Bechaver Collection

The town (Creede) was a close-knit family. I replaced Billy Schultz, who retired from the Game and Fish Department and continued to live in Creede. He was a small man, but large in character as he was greatly respected as both a law officer and private citizen of Creede. I did mostly fish management work and then worked the hunting season. One of the things I loved the most was patrolling the San Juan high country horseback. When winter came we didn't see the bare ground until May. When the snow piled deep I was assigned by my Supervisor, Don Benson, to assist other officers to patrol the San Luis Valley where there was a lot of poaching going on. One night I was riding with Lloyd Hazzard, the Wildlife Conservation Officer in Monte Vista, and WCO Johnny Hobbs of Alamosa, was driving a Dodge sedan. It was a pitch black night and we were watching for spotlighters killing deer on the Trinchera Ranch east of Fort Garland. Lloyd and I saw a light working and started driving without any lights toward the vicinity of the poachers. Unbeknown to us Hobbs had changed locations and was parked in the middle of the same dirt road that we were traveling. Without warning we smashed into the rear of his car. Lloyd and Hobbs weren't hurt too bad. I hit the windshield. I woke up in the back seat of a state patrol car heading for the emergency room at the Alamosa hospital. I was admitted to the hospital as I sustained lacerations to my head and face. I had the neatest impression of a pair of binoculars on my chest for several days. We were lucky our injuries weren't more severe. I have many precious memories of Creede. My first son Brian was born there and now he is the District Wildlife Manager stationed in Saguache for the Division of Wildlife.

Many young people who grew up in a small town knew the local wildlife officer and were attracted to become a wildlife officer. Such was the case for Billy Schultz who influenced Clayton Wetherill to become a wildlife professional. Wetherill said:

> I graduated from Creede Schools in 1953. Dale Hibbs and I were in the same class until his family moved to Salida. I had decided I wanted to be a game warden. Bill Schultz convinced me that I would have to go to college, before I could work for the Game and Fish Department. I joined the Army with George Keller and Denny Ott and when I got out I used the GI Bill to attend Adams State College. The Department wasn't hiring so I majored in education and biology and earned a teaching certificate. I taught school in Kim, Ouray, and Lake City for a few years and I became good friends with Jim Houston, the Wildlife Conservation Officer in Lake City. Jim encouraged me to apply for a Wildlife Conservation Officer position. I was hired and assigned to Walden and later in Sterling before being promoted to Area Supervisor in the San Luis Valley in 1976.

Creede had other native sons who became wildlife professionals. Dale Hibbs spent his early years in Creede but graduated from the Salida school system. He was greatly influenced to become a wildlife biologist by Marvin Smith, the wildlife officer in Salida. Bill Swinehart started his career at the Creede hatchery and became a fish culturist at the Leadville federal hatchery. Leroy Fyock earned his degrees in fisheries and worked in the private sector with major corporations as an environmental biologist and consultant.

Some officers grew up following their fathers and as with many game warden families shared in the work and learned the tricks of the trade. Phil Mason, District Wildlife Manager in Lake City grew up in a game warden family. His father, Allison Mason, was a game warden in Durango. Phil's brother Russ became an officer in the Bailey area. Allison told the author once that he did his best to talk them out of becoming game wardens, but it must have been in their blood. They had a most excellent role model.

Brent Woodward, District Wildlife Manager in Creede, is a third generation wildlife officer. His father Chuck Woodward, an officer in Craig, Colorado followed the steps of his father Walt Woodward, one of the legends in the Division of Wildlife.

The author was introduced to the Rio Grande while on his first horse pack trip with Gene Bassett, the wildlife officer in Bayfield in 1965. Only those who have ridden with Bassett can appreciate what that experience was like. Those were long days with an empty belly and painful saddle sores, but learning to ride and pack horses from such a master was a privilege. That trip was one in which the author first saw the Rio Grande Pyramid and the Ute Lake basin. The rugged beauty of the San Juan mountains attracted him like a magnet. When Ed Bechaver resigned the author transferred from the Cortez district to Creede in the fall of 1966. Billy Schultz became a mentor and a friend. His most memorable advice: "Tear page thirteen out of your law book and then you can't go by the book all the time. But, be sure to put in its place your very best judgment." The author was in awe of the San Juans and the people who live in this isolated and rugged environment. It didn't take long to become comfortable and at home with both.

The life of a wildlife officer can be lonely, but one learns quickly that he or she is not alone. In the author's first winter the Upper Rio Grande Fish and Game Association invited him to its annual meeting where they planned the fish stocking for the following summer. The author learned to stay out of these people's way when they were going to do something for wildlife. The Association did more than stock fish. It printed up thousands of fluorescent orange bumper stickers that said, "God's Wildlife Belongs to All-Don't Waste It." Later they printed posters that were placed in every business and resort, "PROTECT OUR WILDLIFE-Report Wildlife Violators Here". . . and they did as advertised. These were people who cared, listened, spoke out, and took action-what a concept!

One of the enjoyable elements of being a wildlife officer was meeting thousands of good people every year. Many returned each year to their same camping spot, whether it was a resort, campground, or a backcountry site. Over time one learned who camped where, and if they didn't show up, wondered what had happened to them. Hunting camps were warm and friendly. If one arrived at meal time one could be assured of a good meal and sometimes an unusual meal such as squirrel stew, chili that was so hot it couldn't be frozen, or coffee strong enough to float a horseshoe. The fellowship around a campfire meant as much to an officer as it did to the hunters.

Wildlife officers had a status and an aura about them that commanded respect, and young people used to be in awe of a man riding a horse. Punk Cochran impressed one young man more than he'd ever know. Punk was riding his paint mare and leading his faithful mules, Jessie and Whiskey, down the trail past Big Meadows Reservoir where he made a routine stop to check the fishing license of a man and his son. This young man was impressed with Punk's way and perhaps the romantic job, and at that moment he decided he wanted to become a Colorado game warden. Little could Punk have known that in 1988 that young man, Jay Sarason, would become the new wildlife officer in the Creede district and continue the legacy of caring for the wildlife of the Upper Rio Grande.

Punk Cochran transferred from South Fork to Ault, Colorado, in 1969. Mike Zgainer was assigned to South Fork in 1970 and he remembered:

> Punk was like a statue in South Fork. Everyone knew him and respected him. There was Punk's law and way of doing things and the way of the Game, Fish and Parks. It was not easy for a young whippersnapper to take his place. But, I set my own footsteps and did not try and remove such a respected monument.
>
> I immediately faced complaints about not enough fish being stocked in the South Fork and the Rio Grande. Beaver were causing a major problem to ranchers. Punk had also set up Wednesday evening for movie presentations for the Chamber of Commerce in South Fork and there I found myself a target for complaints. I wasn't prepared for being blasted.
>
> The most frustrating things going on were people catching and hogging fish as though they hadn't heard of a possession limit. Many of the summer people were from Texas and were still buying resident licenses. They violated Punk's trust. I tried reasoning with folks and found that only helped a little, but I started to write tickets. Some folks were downright stubborn about admitting they were not eligible for resident licenses or not entitled to ignore the possession limit of ten fish.

Jay Sarason, spoke of the warm satisfaction of working in a small mountain community:

> The first morning I went for a hike up Dry Gulch and I looked around and I was overwhelmed by this awesome country. After being there just a short time I felt I was a part of a community that was family. They appreciated the Division of Wildlife-didn't mean they always agreed, but we were allies for wildlife.

To a new officer every bend in the road and trail is an adventure, but to a two-decade veteran those same routes hold many memories. In the old days game wardens worked all hours of the day and night and seldom took days off. If one saw a vehicle parked in the mouth of a canyon the warden would either be waiting for a "suspect" to return or going for a hike to check out the individual who might be fishing or hunting, or collecting mushrooms or rocks. One never knew for sure what he was getting into. Few wildlife officers carried sidearms. That was part of the Game and Fish Department's culture for many years, but things began to change after Claude Dallas murdered two Idaho Conservation Officers in 1981. That was a wake-up call for wildlife officers nationwide. Within a year the Division of Wildlife intensified officer survival training, including firearms instruction. Each officer had to furnish his or her own firearm, but the Division of Wildlife furnished ammunition and training. Many officers began carrying a sidearm on a regular basis, but to date no officer has fired his weapon in self defense, which is a tribute to their training and professionalism. The Division of Wildlife began furnishing sidearms and other defensive equipment in the mid 1990s.

In the mid-1980s a major change occurred in the way District Wildlife Managers had to work when the Department of Labor, under the Fair Labor Standards Act, sued the Division of Wildlife, because its officers were working too many hours without overtime compensation. The public as well as wildlife officers expected long days, working weekends and holidays, and sometimes not a day off in a month; but that was about to change. The agency did not have the funds to pay for overtime, so it had to restrict the length of time that its officers could work. The District Wildlife Managers fought the lawsuit. How rare in this day for workers to plead to be allowed to work overtime and not be compensated. Traditionally officers worked with little supervision and for as many hours as necessary to get the job done. In the court decision the District Wildlife Managers won some concessions. They were barred from working the extra long hours but were allowed to work a little more than a standard forty-hour week. This meant less time in the public eye and more criticism for not being everywhere at the same time.

Wildlife officers are public servants and in the 1940s the state wanted its game wardens to be more available to the public and went so far as to publish their names, home towns, and phone numbers in the hunting regulations and *Conservation Comments* (precursor to *Colorado Outdoors* magazine). This destroyed their privacy and home life since they received telephone calls at all hours of the day and night, year-round, from people wanting hunting and fishing information. More recently the Division of Wildlife has tried to protect the privacy of officers and their families, to the consternation of some people.

When a new officer arrives in a district he or she immediately inherits the challenges and opportunities of his or her predecessor. When Dave Kenvin left the South Fork district he became the San Luis Valley terrestrial biologist and Jerry Pacheco was assigned to South Fork. Pacheco immediately became involved in game damage issues in the South Fork area and was up to the task of finding new solutions to old problems:

> When I came to South Fork in 1992 the ranchers were really upset about game damage. The 1991 winter brought elk and deer right down onto their ranches between South Fork and Del Norte and it took me a while to meet with all of them, get the paperwork straightened out so they could be paid for the damage that deer and elk had done, get some stack yards built to protect their hay, repair fences, and get them the relief from the damage that they needed. The Division of Wildlife just started a Habitat Protection Program with the ranchers, Forest Service, and Bureau of Land Management. This committee recommends habitat improvement projects on public land that are financed by the Division of Wildlife which enhance deer and elk ranges to hold those animals for a longer period of time so that they don't come down onto private land as much. We have had a few game damage dispersal hunts that ranchers can use to have hunters of their choosing kill a few damage-causing animals and hopefully scare most of them off their land.

The Habitat Protection Program committees work in every basin in western Colorado to address the needs of ranchers and farmers and improve habitat for wildlife as much as possible on public and private land. Cooperation rather than confrontation has been a new concept that is working to perpetuate a way of life that includes wildlife.

Today's wildlife officers spend a lot of their time advising county planners, landowners, and developers about how land-use proposals, timber operations, grazing, travel management plans, subdivision proposals, and other issues might impact wildlife and its habitat. Until recently the District Wildlife Managers were also the field biologists active in fish and wildlife management. Now biologists do the census, habitat evaluation for introductions, species inventories, and season recommendations. Even though the District Wildlife Managers are trained biologists, they are more generalists. Now the field biologists are more specialized and work with the District Wildlife Managers, combining their knowledge and expertise as effective teams.

Law enforcement has continued to be a major responsibility for District Wildlife Managers, but the Fair Labor Standard Act changed the way business was to be done. Officers no longer had the luxury of working overtime. They had to become more efficient. Jay Sarason said:

> When I came on the Creede district I concentrated on those wildlife violations that really hurt the resource such as 'fish hogs' who stockpile too many fish over the bag limit. One doesn't catch these folks by driving around in a state truck in uniform in the middle of the day where everyone sees you. I had to get a little sneaky. I couldn't do it all and I had to concentrate on the most important things. Some people complained that they didn't see me as much as they were used to, but I just had to work differently.

Law enforcement is dependent on four factors. First, a citizenry that is not only law abiding, but also demands compliance with the law by others; second, professional law enforcement personnel that are sensitive to people and proficient in detecting and prosecuting violators; third, a fair and just court system that people know will impose appropriate penalties when necessary; and fourth, good law enforcement programs which educate and inform the citizenry about the who, what, where, when, and why of the law.

In Colorado, prior to 1964, misdemeanor law violations were handled in a Justice of the Peace court. Most of these Justice of the Peace judges were not lawyers by training, but men of common sense and integrity. In 1964 Colorado abolished the Justice of the Peace system. In most cases county judges were appointed who had a legal background. Such was not the case for Robert Wardell, who began his legal profession as a Justice of the Peace in 1960. When he was appointed as County Judge for Mineral County he was one of the few county judges in Colorado who was not a lawyer by training. Yet his many years of experience on the bench made him one of the most respected judges in the San Luis Valley. His common sense approach to justice and sense of fairness punished violators in a way that they accepted, and they respected his judgement. Jay Sarason said of Judge Wardell, "He is the fairest judge I have ever been in court with." Every wildlife officer who has been in his court would probably agree.

The Honorable Robert Wardell, Mineral County Judge. Photo from Robert Wardell Collection

Judge Wardell said of the wildlife officers who brought cases before him, "I found them to be people of integrity, always above board, and very professional. They treated people most fairly."

In western Colorado law enforcement personnel from most jurisdictions work together, because the land is big, rugged, and isolated. Officers train together and are often more than just professional partners, but brothers and sisters behind the badge. The San Juans are one of the most isolated regions in Colorado. The wilderness nature makes it an area where all law enforcement officers must be self sufficient, yet ready to do all that is necessary to help a fellow law officer. As in many western counties, the county sheriffs have been very important partners in wildlife law enforcement. As state peace officers District

Wildlife Managers have also helped the sheriffs with everything from kidnap and murder cases to missing persons. Division of Wildlife pilots have assisted sheriffs for many years to locate lost persons. Gordon Saville had a seventy-five percent success rate at finding lost hunters and hikers in southwestern Colorado during his career. Cooperation between the sheriffs' offices, Forest Service, and the Division of Wildlife in the Upper Rio Grande was a model. During hunting seasons Sheriff Phil Leggitt and his deputies coordinated with District Wildlife Managers and the Forest Service Law Enforcement Specialists to patrol as many roads and back country areas as possible. This cooperation gave each agency more eyes to watch for violations. The Forest Service was notified of road-closure violations and a District Wildlife Manager was notified of any wildlife violations. As a result of a physical presence, friendly and helpful officers gained compliance with regulations.

Violation prevention through information and education is a very important aspect of a good law enforcement program. An important role is fulfilled by the fishing and hunting license agents at various stores and resorts. License agents are often the only contact a sportsperson has with the Division of Wildlife. They pass on their knowledge of fishing and hunting conditions and the laws to the public, and this helps prevent good people from becoming ignorant violators. License agents for the most part have been very intolerant of people not buying licenses. Some people thought it was an accident that a wildlife officer caught them without a license, when they were an easy target because a telephone call tip brought the law down upon them. This intolerance for disrespect of the law and wildlife has been so evident that visitors soon learn that in order to fully relax and enjoy hunting or fishing one must also obey the law.

As the reader has witnessed throughout this book, citizens have participated in the protection, recovery, and management of wildlife in the San Juans for more than 100 years. They too are some of the stewards of the wildlife. The job of perpetuating wildlife cannot be done without citizen volunteers. Why do some people work so hard to protect wildlife and wild places? There are many reasons, but perhaps they are all in love with wildness itself-that characteristic which draws people to these pristine mountains, forests, and waters. Some people want to give something back to these mountains for all the life changing experiences they have provided. People take care of what they cherish, and perhaps this is another reason for wildlife having such a high value. Wildlife stirs a sense of adventure, excitement, and compassion in people who are involved with it. But there is a danger of loving nature to death, if only emotion stirs our hearts to action. Susan Dieterich, Wildlife Rehabilitator, used the term "pragmatic compassion" in a presentation to describe actions that are sometimes difficult and even politically incorrect, but are done for the long term benefit of wildlife-such as allowing an animal to die a natural death. Sometimes our humanizing wildlife does get in the way of sound decision making. It doesn't mean we love wildlife less; it means that there are times when we must let nature take its course, as cruel as it may seem.

There are some who would abolish wildlife and land management altogether and "let nature take its course." We have so spoiled the natural systems that to turn our back and not be responsible for the natural environment would be devastating to wildlife. Society isn't ready to accept the natural consequences of widespread starvation of elk and deer in severe winters if herds were allowed to increase and not be managed. Taken to the extreme such an idealistic position would require that humans no longer inhabit or visit the San Juans or most of Colorado and even the Rocky Mountains. The future of wildlife, and indeed our own survival, depends on community values that have a conservation ethic toward soil, plants, water, air, and space. The answer is not in totally prohibiting activities such as grazing, logging, fishing, hunting, hiking, road and home building, but learning to live and recreate upon the land-gently.

Stewardship is the responsibility of whole communities, not just local, state, and federal agencies. Governmental agencies are merely mechanisms that society uses to manage resources. Nature has no respect for and has reason to fear the democratic process. The mountains can't vote for their future nor do they assign their proxies to any representative as some would claim. Historically, the majority opinion has often erred in the way it has demanded resources be managed. In the quest to live off the resources of these mountains, political decisions often directed what was done to the mountain. They were based on economic values that have almost always outweighed the needs of the mountain. The mountain knows its needs and processes, but mankind is a long way from fully understanding them.

Even though there are some places in southwestern Colorado which are some of the last strongholds of real wildness in Colorado, many others have already been lost. Over-managed forests and rangelands have lost their wildness. Abused land erodes its soil into streams and chokes pristine waters and degrades delicate ecosystems. Noisy people disrupt the quiet solitude of a clear blue mountain lake. To some the San Juans and its wildlife are here merely for entertainment or economic gain, but this magnificent place is not an amusement park or a zoo. It is a living ecosystem that humans are a part of and must be responsible for.

Selfishness and greed are human traits which have devastated land, water, air, and resources that societies depend upon. One can only hope that people will act responsibly and respectfully toward what our Creator has given us. What a tragedy it would be if people ruin the very character of this land that attracted them here in the first place. Our wildlife heritage has recovered from near extinction over the past 100 years, but what will the new millennium hold for wildlife? It will be good if we learn to live gently upon the land.

Wild places and wildlife call to us in languages that words alone cannot articulate. Next time you stand in a forest or see a deer in a meadow, listen carefully to the call of the wild. Perhaps you can hear it saying: "Enjoy Colorado's wildlife heritage and carry on your legacy of stewardship."

APPENDIX

Upper Rio Grande Wildlife Officers

The personnel assignments and dates of service of the early game wardens are incomplete. In the early days these game wardens worked the whole San Luis Valley:

F.B. Orman, Chief Game Warden, Pueblo 1901—most of southern and southwestern Colorado
W.D. Wilson, Chief Game Warden, Creede, 1903 to 1908—most of southwestern Colorado
Joseph H. Whiteley, Chief Game Warden, Creede, 1909—most of southwestern Colorado
Walter Cambell, Chief Game Warden, Alamosa, 1912—primarily San Luis Valley
Frank Sabine, Chief Deputy Game Warden, Alamosa, 1915—primarily San Luis Valley
Clarence Goad, Chief Deputy Game Warden, Alamosa, 1918 to 1943—primarily San Luis Valley
Bill Krebbs, Deputy Game Warden, Del Norte, 1930s and early 40s (dates uncertain)

Creede District:

Bill Shultz, Game Warden/Wildlife Conservation Officer, 1946 to 1965
Ed Bechaver, Wildlife Conservation Officer, 1965 to 1966
Glen Hinshaw, Wildlife Conservation Officer/District Wildlife Manager, 1966 to 1988
Jay Sarason, District Wildlife Manager, 1988 to 1993
Tammy (Fox) Spezze, District Wildlife Manager, 1993 to 1994
Brent Woodward, District Wildlife Manager, 1994 -

South Fork District:

Earl "Punk" Cochran, Game Warden/Wildlife Conservation Officer, 1946 to 1970
Mike Zgainer, Wildlife Conservation Officer/District Wildlife Manager, 1970 to 1979
Dave Kenvin, District Wildlife Manager, 1979 to 1992
Jerry Pacheco, District Wildlife Manager, 1992 -

San Luis Valley Supervisors:
Earl Downer, ? to ?*
Clyde Slonaker, ? to 1960*
Jack Andrews, 1960 to 1961*
Don Benson, 1962 to 1968
Harvey Bray, 1968 to 1970
Walter Shuett, 1971 to 1991
Clayton Wetherill, 1976 to 1991
Jerry Apker 1991, -
* Supervised the law enforcement and trappers in the San Luis Valley

San Luis Valley Wildlife Technicians who worked in the Upper Rio Grande:

Ron Desilet	Howard Spear	Tom Martin	Clete Nolde
Bob Rouse	Ed Dumph	Jeff Johnson	Ron Rivale

Area Biologists who worked in the Upper Rio Grande:

Mike Japhet	John Alves	Beverely Motz
Kirk Navo	Dave Kenvin	Don Masden

The Administrative Assistant who holds them altogether:
Karen Rhoads

Forest Rangers of the Upper Rio Grande

There were two ranger districts. One was the Pyramid District and the other was the Creede District. They were consolidated in 1952. In 1994 the Creede District was consolidated with the Del Norte District to become the Divide Ranger District headquartered in Del Norte, Colorado.

Pyramid District Rangers

(headquartered at the present site of the Bristolview Guard Station)

Name	**Date of Service**
Roderick Day	1907 to 1908
Martin Willy	1907 to 1908
Glen K. Munsell	1908 to 1910
Don C. LaFont	1910 to 1919
Edwin Bennett	1919 to 1922
Elbert E. Vanaken	1922 to 1934
Fay Franklin	1935 to 1943
Weslet R. Haynes	1943 to 1944
Robert L. Dunstan	1944 to 1947
Waldemar Winkler	1947 to 1952

Creede District Rangers

Name	Date of Service
Clarence Bradt	1906 to 1908
Roderick Day	1907 to 1908
W.F. Bowers	1908 to 1909
Francis F. Joy	1908 to 1909
Glen.K. Munsell	1910 to 1911
Reed Lee	1912 to 1913
S.E. Doering	1913 to 1920
D. M. Truman	1920 to 1926
Fay Franklin	1926 to 1935
Donald E. Kipp	1935 to 1941
P.L. Heaton	1941 to 1943
Fay Franklin	1943 to 1951
Carl F. Henderson	1952 to 1957
John Minnow	1958 to 1961
Roy C. Kuehner	1961 to 1965
Bob Pizel	1965 to 1973
Gary McCoy	1973 to 1978
David Nichols	1978 to 1979
Larry Robinson	1979 to 1985
Dick Lindenmuth	1985 to 1989
Carlos Pinto	1989 to 1994
Julie Howard	1994 to

Creede Federal Fish Hatchery Superintendents

The Creede Fish Hatchery was a substation of the Leadville Hatchery. The position of Superintendent was in Leadville and the Creede substation had a Manager or Foreman. Without getting into bureacratic titles, the following men were in charge of the Creede Hatchery:

Superintendent	**Tenure**
Charles Fuqua	1929-1933
John Harrington	dates uncertain
Gene Mason	1933-1938
John Thompson	1938-1946
Tom French	1946-1954
Norman Wilkerson	1954-1960
Jerry Pearson	1960-1965

San Luis Valley Wildlife Commissioners

The Governor appoints citizen Game and Fish Commissioners that must be approved by the Senate. A political balance between Republicans and Democrats has been required by law. Commissioners are from many backgrounds. The commission has varied from six to ten members to represent each area of the state.

Commissioner	**Term**
Otis McIntyre	1937-1947
Theodore Eckles	1947-1953
Henry Lague	1953-1959
Parker Sooter	1959-1965
Floyd Getz	1965-1971
Lonnie Pippin	1971-1975
Dr. Jay Childress	1972-1979
Donald Fernandez	1979-1987
Larry Wright	1987-1992
Arnold Salazar	1992-

BIBLIOGRAPHY

Alves, John. "Status of Rio Grande Cutthroat Trout in Colorado," Colorado Division of Wildlife, March 30, 1998.

Barrows, Peter and Holmes, Judith. *Colorado's Wildlife Story,* Colorado Division of Wildlife, 1990.

Boyd, Raymond J. "Rio Grande Elk Study," Colorado Game and Fish Dept.,1965.

DuBois, Coert. "Report on the Proposed San Juan Forest Reserve, Colorado," U.S. Forest Service Archives, 1903.

Bryne, G. and V. Schwieker. "Fisher, Lynx, Wolverine Observations and Records for Colorado," Colorado Division of Wildlife, 1998, Unpublished Report.

Colorado Conservation Comments, Vol. I, Number 1, Joyce Becker Collection.

Colorado General Laws, Chapter XLI, 1877, game pages 483-486.

Couses, Elliott. Ed. *The Journal of Jacob Fowler,* 1821-1822, University of Nebraska Press, Lincoln, 1970.

Crofutt. *Crofutt's Grip-Sack Guide of Colorado,* Vol II, 1885.

Ellis, Max M. *The Fishes of Colorado,* The University of Colorado Studies, Vol XI, 1914.

Fentzlaff, Richard. "The Ute Hunting Rights," *Colorado Outdoors,* Sept-Oct. 1981, page 28-30.

Flader, Susan L. *Thinking Like a Mountain,*University of Missouri Press, Columbia, Missouri, 1974.

Fraser, Walter. "Fish and Game Commission Report to the Governor," 1915.

French, Tom A. "Spawn from Lake San Cristobal," *Progressive Fish Culturist,* Vol 10, 1948

Game Fish and Parks Division. *A Look Back, a 75 year history of the Colorado Game Fish and Parks Division,* 1972.

Grant, Blanche C. *When Old Trails Were New,* The Rio Grande Press, Inc., Chicago, page 344.

Griswold, P.R. *San Luis Valley Historian,* San Luis Valley Historical Society. Number 4, 1988, pages 1, 12.

Howard, George. Personal Journal, Montrose County Historical Museum, 1872.

Kindquist, Cathy E. *Stony Pass, The Tumbling and Impetuous Trail,* San Juan County Book Co., Silverton, Colorado 1987.

Jacobs, Janis. *Ribs of Silver, Hearts of Gold: The Story of Homesteading, Ranching, and Dude Ranching,* B & B Printers, Gunnison, Colorado, 1994.

LaFont, John. *The Homesteaders of the Upper Rio Grande,* Oxmoor Press, Birmingham, Alabama, 1971.

Lorbiecki, Marybeth. *A Fierce Green Fire (A Biography of Aldo Leopold),* Falcon Publishing Co., Helena, Montana,1996.

Marsh, Charles S. *The Utes of Colorado: People of the Shining Mountains,* Pruett Publishing Co. Boulder, Colorado1982.

Mills Annotated Statutes of the State of Colorado. Vol 3, 1891-1895, pages 616-617.

Munsell, Glen. Personal Diary, 1908-1911, Dorothy & Charles Steele Collection.

Newton, JoRene. *Ribs of Silver, Hearts of Gold: A History of the Natural Resources of Mineral County and the Upper Rio Grande,* B&B Printers, Gunnison, Colorado, 1995.

Nossaman, Allen. *Many More Mountains, A History of Silverton and Mining in the San Juans,* Sundance Publishing, Denver, Colorado, 1993 Vol. 1 1989, Vol 2, Vol 3, 1999.

Nordenskiold, G. *The Cliff Dwellers of Mesa Verde,* 1893, page 12.

North, Oliver. *The Field, The Country Gentelman's Newspaper,* May 4, 1878.

Oliver, Martha, "Cattle in the Valley," *San Luis Valley Historian,* Vol 17, No. 4, 1985.

Olterman, Kenvin and Kufeld. "Moose Transplant to Southwestern Colorado," Alces, Vol 30, 1994.

Peterson, David. *Ghost Grizzlies,* Johnson Publishing, Boulder, Colorado, 1998.

Pettit, Jan. *UTES, The Mountain People,* Pruett Publishing, Boulder. Colorado, 1982.

Reyher, Ken. *Antoine Robidoux and Fort Uncompahgre, A Story of the Fur Trade in Western Colorado,* Western Relections, Ouray, Colorado, 1998.

Rickard, T.A. *Across the San Juan Mountains 1902,* Bear Creek Publishing, Ouray, Colorado, 1980.

Ryder, Ronald A. *A Checklist of the Birds of the Rio Grande Drainage of*

Southern Colorado, Colorado State University, July, 1965
San Luis Valley Fair, Inc. *Folks & Fortunes, Saga of the San Luis Valley in Colorado,* Vol 1, Number 1, 1949.
Sheldon, Mitchel G. "Silvertip Search," *Colorado Outdoors,* Jan/Feb 1956, pages 28-29.
Smith, P. David. *Ouray, Chief of the Utes,* Wayfinder Press, Ridgway, Colorado, 1990.
Swift, Lloyd W. "A Partial History of the Elk Herds of Colorado," *Journal of Mammology,* 26(2): pages.114-119, 1945.
Tischbein, Geoff. "Are There Still Grizzlies in Colorado?," *Colorado Outdoors,* January-February, 1980, pages 12.
United States Department of Agriculture. *Work Plans and Annual Reports 1924-1945, Rio Grande National Forest, Appendix.*
Wason, Henry. *The Indian Fish and Game,* Creede, Colorado, 1926.
Wilkerson, Norman G., Jr. "Transporting Small Live Trout in Sealed Polyethylene Bottles," *The Progressive Fish Culturist,* Vol 20, No 4, October, 1958, page. 175.
Wiltzius, William J. *Fish Culture and Stocking in Colorado, 1872-1978,* Colorado Division of Wildlife, 1985.
Southern Ute Homepage. Internet: www.southern-ute.nsn.us/history/index.html

INDEX

ABOUT THE AUTHOR

Glen Hinshaw was born and raised in Denver, Colorado. Glen grew up with a fishing pole in his hand, following his father to the wildest places in Colorado to find those secret fishing spots. While at North High School he set his goal to become a Game Warden. He earned a Bachelor of Science degree in Wildlife Management at Colorado State University in 1963. He went to work for the Colorado Game and Fish Department in 1963 and was stationed in Cortez, Colorado. In 1966 he transferred to the Creede District. While in Creede he was recognized for his work and received the Division's "Officer of the Year" award in 1968. In 1985 he was named the Trout Unlimited "Conservationist of the Year" for his work to improve fishing in the Upper Rio Grande. In addition to performing all the duties of a Wildlife Officer while in Creede, he also participated in many aspects of community life being one of the first Emergency Medical Technicians, serving on the first Mineral County Ambulance Board, and as first chairman of the San Luis Valley Emergency Medical Services Council. He was active in school issues by serving as chairman of the first School Improvement Committee and being an assistant wrestling coach. He taught Bible classes and sometimes filled the pulpits at the Community and Baptist churches. He was one of the founders of the San Luis Valley Underwater Recovery Team. Glen was on many search and rescue missions while serving as the wildlife officer in Creede. He wrote a weekly newspaper column in the Creede Candle to keep people informed about wildlife issues. He was promoted to Information and Education Specialist in Montrose in 1988 and while there developed several education programs for the Division of Wildlife. In 1997 he received one of the highest awards in Colorado from the Colorado Alliance for Environmental Education: "The Enos Mills Lifetime Achievement Award for Environmental Education." At retirement in January of 1998 he was the Division of Wildlife's Education Coordinator for Western Colorado. He is a man in love with people, wildlife, and the wild places of the San Juan Mountains. He lives in Montrose with his wife Carol.